T0297253

PURE CULTURES OF ALGAE

M. W. BEIJERINCK

PURE CULTURES

OF

ALGAE

Their Preparation & Maintenance

BY

E. G. PRINGSHEIM

Department of Botany, Queen Mary College,
University of London

CAMBRIDGE

AT THE UNIVERSITY PRESS

1946

CAMBRIDGE
UNIVERSITY PRESS

University Printing House, Cambridge CB2 8BS, United Kingdom

Cambridge University Press is part of the University of Cambridge.

It furthers the University's mission by disseminating knowledge in the pursuit of education, learning and research at the highest international levels of excellence.

www.cambridge.org
Information on this title: www.cambridge.org/9781316613207

© Cambridge University Press 1946

First published 1946
First paperback edition 2016

A catalogue record for this publication is available from the British Library

ISBN 978-1-316-61320-7 Paperback

A NOTE ON THE FRONTISPIECE

In selecting M. W. Beijerinck's *portrait as the frontispiece of this book, the author has been guided not only by the consideration that* Beijerinck *was the first to apply the technique of bacteriology in the preparation of pure cultures of algae, but that, in company with* Pasteur *and* Winogradsky, Beijerinck *also laid the foundations of General Microbiology.*

The present author has, throughout his career, been greatly influenced by Beijerinck's *practice of securing enrichment of specially adapted ecological types by a consideration of their biological properties, thus achieving what he called 'natural pure cultures'. Such cultures have proved very helpful in elucidating the physiological interrelation between algae and other micro-organisms, without a knowledge of which we cannot hope to understand their conditions of life or proceed methodically to obtain pure cultures, which must form the basis of exact biochemical research and of the investigation of the physiology of development.*

CONTENTS

FOREWORD

In many branches of biological science progress is dependent on the gradual improvement of the methods of investigation. A methodical study of the enormous diversity of simple forms of life, in part differing among one another only in slight and rather elusive particulars, is only possible if extensive and homogeneous material becomes available. Attempts to obtain such material by artificial means, based on the technique developed for bacteria and later for fungi, were therefore instituted already in the last century, and Pringsheim rightly hails Beijerinck as a pioneer in this direction. Slightly later a vigorous development of the culture technique followed at Geneva and, under the leadership of Robert Chodat, numerous green algae were grown in this way. The remarkable polymorphism, and the bizarre forms often appearing in his and his collaborator's cultures, however, gave rise to an element of scepticism as regards the value of such cultures, which found expression in various ways.[*]

Meanwhile, however, the technique of pure culture was being investigated by the author of this book from a new angle. Recognizing the diversity of ecological factors to which the organisms concerned are subjected, Pringsheim developed a more elastic method of approach and endeavoured to mould his culture technique to fit the needs of the individual alga or flagellate. How great a degree of success has thus been achieved will be evident to anyone who has made use of the numerous cultures he has now at his disposal. These cultures are healthy populations composed of single organisms, and in general display none of that extraordinary

[*] Cf. e.g. Fritsch, Presidential Address, Brit. Assoc., Leeds, 1927, Sect. K, p. 15.

variability which made one feel dubious as to the value of the cultures grown by the Geneva school.

Moreover, Pringsheim has recognized the fundamental fact that, unless required for special purposes, in particular for physiological investigations, it is unnecessary to have bacteria-free cultures. His soil- and water-culture method, offering as it does scope for unlimited variation in the cultural conditions, marks in my opinion one of the biggest contributions he has made to the methodology of the culture of lower organisms. Since in nature Algae and Flagellata are always associated with bacteria, it is probable that the latter may often play an essential role in relation to the life processes of the former. The absence of such cooperation in bacteria-free cultures may often be one of the prime causes of failure to secure them, although it is clear that with many lower organisms growth in the absence of bacteria can be successfully achieved.

The author himself enumerates towards the end of this book the many diverse purposes for which unialgal or bacteria-free cultures can be utilized. It cannot be doubted that genetically homogeneous material of the simpler forms of life will be used to an increasing extent in the study of fundamental physiological principles. At the same time the perfection of methods of culture will render feasible a more direct approach to some of the problems of fresh-water biology than has hitherto been possible. A reassembly of mixed populations, both of microscopic plants and of animals grown individually in pure culture, should render possible a clearer elucidation of the problems of competition under standardized conditions and of the mutual interreactions of the components of a population upon one another. This may well indeed constitute a major step in the elucidation of the first links in the food chain.

The possibility of rearing cultures from a large diversity of lower organisms entails the risk that this may come to be

an end in itself, and that many of the forms involved become
known to science only in this condition, without definite information as to their mode of occurrence in nature.* The
pursuit of inquiry along these lines will often provide data
of appreciable morphological or physiological interest and
may widen our taxonomic outlook, but, until the role and
mode of occurrence of such forms in nature have been established, the information gleaned from their culture is largely
academic and is of practically no value to the ecologist.
Pringsheim therefore rightly emphasizes the importance of
a preliminary study of all organisms taken into culture, both
with a view to ascertaining their conditions of existence and
to establishing their mode of occurrence in nature.

Any scientific method depends for its success on attention
to a multitude of detail and requires the exercise of considerable patience on the part of the investigator until the
requisite manipulative skill has been acquired. It is impossible to overestimate the importance of data supplied as
a result of experience accumulated over many years by an
expert like Pringsheim, and this volume should go far in the
direction of initiating the novice into the methods of pure
culture and of fostering interest in this important technique.
If it achieves the success which it deserves, it may well open
up a new era in the intensive investigation of the many
aspects of lower organisms that claim the immediate attention
of biologists. In many ways the study of this branch of
microbiology is of outstanding economic importance, quite
apart from its fundamental interest in exposing the characteristics and modes of life of lower forms of plant and animal
organization.

F. E. FRITSCH

Cambridge
June 1944

* Cf. for instance the large number of new Heterokontae described
by Pascher in vol. XI (1937) of Rabenhorst's *Kryptogamenflora*.

PREFACE

Since I described the methods of working with cultures of algae used in my laboratory at Prague (1926 *a*) there has been no really useful publication in this field of biology until Bold (1942) gave a review based on literature* and supplemented it with valuable comments. In the meantime, however, the technique has been further developed, so that it is now easier to prepare pure cultures, and the method has been extended to embrace more species.

On the other hand, it is becoming increasingly obvious that bacteria-free cultures are needed, not only for experiments on nutrition, but also for certain problems of ecology, hydrobiology and fishery which cannot be effectively approached without such cultures.† Many problems relating to propagation, heredity and genetics cannot, moreover, be solved without the help of at least unialgal cultures. Even the original aim of algal study, namely, that of describing species, cannot, as we have been forced to acknowledge, always be attained without cultures starting from single cells. For these reasons almost all workers on algae will in future be obliged to use cultural methods.

My object is to show that it is by no means difficult to acquire the necessary skill. With some patience and zeal, valuable results can be obtained after a relatively short period of training in this attractive field of biology.

---

* Bold's paper gives an almost complete list of relevant references.

† I do not wish to give the impression that I regard pure cultures as indispensable for ecological work in the field. Such over-emphasis could only be harmful to the adoption of the technique of pure cultures. Simple cultural methods will, however, prove very helpful to workers in this field, and investigations based on pure cultures will afford a foundation for future floristic and applied work on algae.

By publishing an account of the practical conclusions reached during many years of experimentation with algae, I hope to draw the attention of biologists to the great possibilities opened up by the culture of these organisms. Existing circumstances have not for some time afforded me the opportunity of training junior research workers, which would have been a much more effective method of imparting instruction than a written account can ever be. If fortune favours me, I hope to be able to act in this capacity again. The progress of science depends not only on research, but also on tradition. Just as strains maintained in my culture collection have been available to everyone, I would like to give all who are interested the opportunity of avoiding my mistakes and of profiting by my experience in this field of biology. I still hope to find the means for the establishment of a small station, devoted to research on algae.

For those who question the value of such an institution the following considerations are relevant. Much time and labour have been wasted by the failure to define and recognize species of algae and flagellates. Much physiological research has been invalidated by the use of impure or unidentified algae or could not be undertaken for lack of suitable material. Many morphological investigations have been interrupted or abandoned because living material was no longer available. Many theoretical problems are best solved with the help of these small organisms. Test objects suitable for medical and other purposes could probably be found among the cultivated species. Little is known about vitamin production, about the sources of food for animals of vital interest to mankind, about the origin of organic substances in sea and fresh water, about the relation of soil algae to fertility, and so on. These problems are likely to be solved with the help of pure or combined mass cultures and their biological and chemical investigation. Nobody should therefore overlook the probability that we are at the threshold of a scientific edifice which must be built sooner or later, comprising the specific ecology and physiology of organisms of which only morphological data are as yet known.

Many of the experiments and observations which form the basis of the experience detailed in the following pages are the fruits of the years spent in England after I had been deprived of my working place in Prague. I find it difficult to express my indebtedness to those who made it possible for me to continue my researches, but special mention should be made of the Society for the Protection of Science and Learning, the Royal Society of London, the British Association for the Advancement of Science, and the University of London, who assisted me with grants. Many thanks are due to Queen Mary College, University of London, and especially to the head of its Department of Botany, Prof. F. E. Fritsch, who did all in his power to improve the conditions of work, and who encouraged me much by his friendly interest; as well as to the staff of the Botany School in the University of Cambridge, where I carried on my work after Queen Mary College was evacuated from London. Without the help of my wife, my only assistant, I could not have preserved the culture collection on which my investigations are based.

E. G. P.

CHAPTER I

INTRODUCTION

The study of algae has entered on a new phase through the extensive utilization of cultures for various purposes. As was the case in bacteriology and mycology, a much greater development of this mode of investigation may be expected in the near future. Progress in the study of algae will largely depend on the use of the existing methods of culture and their successful modification for special purposes.

The technique of obtaining pure cultures of algae has been considerably simplified and improved during the last few years. There is actually nothing new in the methods now adopted, but a few devices and accumulated experience render success more likely than it was ten years ago and result in a saving of time and material.

For the better understanding of the nature of the problems which can already be solved and to do justice to the scientists who prepared the way, the historical development will be traced.

Famintzin (1871) was probably the first to emphasize the possibility of ascertaining the nutritive needs of an alga with the aid of solutions of inorganic salts, a line of investigation pursued by Molisch (1895, 1896) and Benecke (1898) with considerable success. As a result the correspondence, in this respect, between algae and vascular plants was recognized, except that some of the former may not need calcium. It was only later realized that additional chemical elements (e.g. manganese) are necessary for algae as for flowering plants in addition to the ten long claimed as sufficient. But up to the present our knowledge on this point is far from satisfactory.

Miquel (1890–92, etc.) developed devices for cultivating diatoms in artificial media without securing pure cultures devoid of bacteria, which were first claimed by Richter (1903) and Chodat (1904).

The first step in this direction with respect to other algae was taken by Beijerinck (1890, 1893), who, adopting the technique devised by Robert Koch for bacteria ten years earlier, used gelatine for fixing germs at definite places. Klebs (1896, p. 184), however, doubts whether Beijerinck's cultures were really free from bacteria. Since the publication of Beijerinck's classical paper the methods of growing algae have undergone much change. Gelatine has proved unsatisfactory, since it is readily disintegrated by bacteria, which are always far more numerous than algal cells. The introduction of agar-agar instead of gelatine by Tischutkin (1897) marked an important step forward. A similar method was proposed by Marshall Ward (1899), who did not know of the work of Tischutkin and did not undertake any investigations as the latter did. It is unfortunate that Tischutkin's paper remained largely unknown, since it could have accelerated progress in culturing algae.

Chodat (1900, 1904, 1909, 1913) also used agar, but employed Erlenmeyer flasks instead of the Petri dishes used by Tischutkin and Marshall Ward. This complicated the technique without contributing any advantage. His method has even been used as late as 1926 by Bristol-Roach in her otherwise very able experiments. As recently as 1932 Skinner allowed the agar medium, with contained algal cells, to solidify in test-tubes, which must be broken to remove the algae. The writer has returned to Tischutkin's method (Pringsheim, 1912 and later), which has been further developed as described in the subsequent pages.

Another line of development originated from Klebs's often cited *Bedingungen der Fortpflanzung bei einigen Algen und*

Pilzen (1896). He made no effort to obtain bacteria-free cultures, because he did not believe that this could be achieved or, at any rate, not without great trouble. He therefore used large species which could be handled by simple methods, and cultivated them as far as possible in media composed solely of water and inorganic salts. It is now known that he underestimated the part played by micro-organisms in the results thus obtained, although some of these were of the greatest importance for further progress.

An important achievement of Klebs's was the use of zoospores as the starting-point for cultures of a single species. Immediately after their release from the parent cells these, as we now know, are free from contamination by micro-organisms, especially bacteria, which usually adhere to the surface of an alga. Zoospores are often the only means of obtaining pure cultures. Klebs defined the conditions for the formation of zoospores, to which Gerneck (1907) added further data. The latter was the first to rear bacteria-free cultures from single zoospores. Like his teacher Chodat (1909) and like Beijerinck, he worked with small species of algae, which can be treated almost like bacteria and yeast, the method of cultivation of which was already fairly well known through the researches of L. Pasteur, R. Koch and E. Ch. Hansen. This choice of small species, which would grow in the media used by Chodat and his collaborators, restricted the selection to those which are adapted to a habitat rich in organic compounds. This restriction was more extreme than was realized at the time, and exaggerated conclusions were deduced from the results. Furthermore, the conditions in such cultures contained within a compact block of agar were far more unfavourable than was at first apparent, so that only few species could withstand them. Finally, the conditions in Chodat's cultures were so different from those in nature that the cells were morphologically abnormal.

Despite these restrictions the experimental work of Chodat's Geneva school proved to be very noteworthy and even invaluable.

The first to review the whole situation was Küster (1907), whose book on the technique of the culture of microorganisms contains a very valuable account of the different media and devices used in experimental work on algae, as well as of the influence of external conditions on their morphology and reproduction. In the later editions, however, the author's lack of detailed familiarity with this field of research makes itself felt. Moreover, the Klebs school to which he belonged had not properly appreciated that bacteria-free cultures were essential for many of the problems to be investigated.

Meanwhile, a different method of isolating single cells for pure cultures was gradually developed. Klebs (1896, p. 184) had already picked up zoospores and flagellates with the help of capillary pipettes. Zumstein (1900) was apparently the first to obtain bacteria-free cultures by means of this technique, which he combined with another means of getting rid of contaminating bacteria, namely, making the medium as acid as was compatible with the growth of the alga concerned. This of course was only possible with acid-tolerant forms, such as *Euglena gracilis*, which Zumstein was cultivating. The same combination of methods was used by Ternetz (1912) and by Pringsheim (1921 b, 1934 a) in dealing with *Euglena gracilis*, *Astasia ocellata* and *Chilomonas paramecium*.

The pipette method was first recommended by Pringsheim (1921 a, p. 402) and improved by Lwoff (1923, 1929). By isolating single cells in a sterile medium and repeating this procedure many times the ratio of number of bacterial to algal cells is gradually reduced until individual cells of the alga could be transferred to a sterile medium and allowed to multiply.

Similar methods had formerly been used for ciliates. The pipette method thus developed into the washing method.

Further improvement of the cultural technique was achieved by introducing several minor but useful devices, and by applying the experience that had meanwhile been gained as to the factors necessary for the existence of forms showing special adaptations or specially sensitive to changes in the environment. The work of Bouilhac (1897), Chodat (1904) and Pringsheim (1913) led progressively to the attainment of bacteria-free cultures of Cyanophyceae, the last-named undertaking with such cultures a study of the physiology of their nutrition. Pringsheim (1912, 1918) obtained the first healthy cultures of Conjugales, which his pupil Czurda (1926) succeeded in growing on agar free from bacteria. The first colourless alga to be grown in pure culture was *Polytoma* (Jacobsen, 1910), whilst Pringsheim (1920, 1921 *b*) was able to improve the method and to use it on a larger scale through the discovery that these and other forms were acetate organisms (Pringsheim, 1921 *b*, 1935).

Organic compounds were first used in algal cultures by Bouilhac (1897, 1898) and Treboux (1905), who employed sugars and organic salts respectively. The first more elaborate study as to the food value of various organic substances was probably that of Matruchot and Molliard (1902) on *Stichococcus*.

As a step towards the understanding of the ecology of algae, an experiment first carried out by Beijerinck (1901) is important. He found that certain Cyanophyceae, such as species of *Nostoc*, *Anabaena* and *Cylindrospermum*, usually develop when small quantities of soil are covered with a relatively large amount of water, only enriched by a low concentration of phosphate. He called these Nostocaceae oligonitrophilic and suspected them of utilizing atmospheric nitrogen, a conjecture which has since proved correct. When larger amounts of soil or complete nutrient solutions are employed, Oscillatoriae, diatoms or other algae are mostly stimulated to active multiplication.

Jacobsen (1910), on the other hand, demonstrated the effect of the addition of a rich supply of organic substances to cultures of soil algae. Volvocales, such as *Carteria, Chlamydomonas* and *Sphondylomorum*, showed vigorous development in the presence of protein if the cultures were illuminated, while in the dark *Polytoma* appeared. It was on the basis of such cultures that Pringsheim (1921 *b*, 1935) built up his theory of acetate organisms.

In relation to the improvement of algal media, the utilization of soil extracts (Pringsheim, 1912, p. 326; 1936 *a*; Mainx, 1927 *b*, p. 323), the effective components of which are probably soluble iron compounds, and of peat extract (Wettstein, 1921), which is valuable for its acidity and low salt concentration, is noteworthy. These substances can be used for regulating and buffering the *p*H of the medium. Uspenski (1925, 1927) has made a special investigation of the influence of iron supply in algal cultures; this has been supplemented by Pringsheim (1930, 1936 *a*).

If algae are to be kept under constant conditions or to be maintained for more than a few months, daylight is insufficient, because it is too variable and, during the winter, too feeble. Hartmann (1921) remedied the defects of electric illumination by allowing the rays to pass through a screen of cold water, thus eliminating the infra-red radiation which injures the algae by heating the cultures and is also harmful through its desiccating effect on the media. Pringsheim (1926 *a*, p. 293), in summarizing his own experience and that of his collaborators, described an improved model of the 'artificial sun' of Hartmann.

During recent years other authors have endeavoured to summarize, wholly or in part, the methods used for algal culture and the results obtained. Kufferath's book (1930) is of little value, but Vischer's instructions (1937) for the culture of Heterokontae might serve as a model for such treatises. The papers of Lwoff (1932), Provasoli (1937–8) and Hall (1939) on the nutritional needs of algae and Flagellata supply a useful basis on which the future science of algal nutrition can be built.

CHAPTER II

SELECTION AND PREPARATION

1. CHOICE OF MATERIAL

When selecting algal material to be raised for a morphological or physiological study, one may proceed in various ways. A start may be made by finding out what will grow from among a mixture of forms, such as are found in a pool, ditch, etc.; or a search could be made for a population practically composed of one species. Both methods may give valuable results, but both tend to afford only such species as will readily grow under certain conditions.

For several years now I have adopted a third procedure, which affords most interesting results and consists in picking out single specimens from a mixture, with a view to analysing the algal flora of a certain habitat or to the isolation of certain species belonging to a definite taxonomic group (cf. also Chu, 1942). In this way, even in a small pond, a multitude of species may be found, the existence of most of which is usually unnoticed owing to the competition of others and the lack of food for all.

This can be demonstrated by subjecting a natural mixture of species, such as would be present in nature, to various conditions. If, for example, we take a seemingly barren mud from the bank of a river and add to different portions water only, Beijerinck's solution, and water with a small piece of cheese respectively, we should find after some weeks practically no growth in the first, a few filamentous and unicellular species in the second, and a rich and varied flora comprising some unexpected species in the third.

In another set of experiments mixed material from a small

garden pond was analysed by subjecting it to varying treatment. Some thirty species of green and unpigmented algae were isolated, most, of which had not been recognized by mere microscopic investigation. Certainly more than double the number would have been found, if more time could have been spent on the task.

When analysing an algal flora, certain forms are found not to grow well in the usual inorganic media, even when various concentrations and pH values are employed, while others will not grow at all in them. When that is so, the soil-and-water culture method (p. 16), first devised for cultivating heterotrophic flagellates, is useful.

2. PREPARATORY CULTURES

It is often advisable to undertake preparatory cultures before attempting pure ones. There are manifold reasons for this procedure:

By cultivating a species for some time under various circumstances, we gain experience concerning the conditions suitable for its existence and multiplication.

We can pick up single cells and allow them to grow, independently of other organisms which might obscure the results. We are thus able to start purification with healthy and relatively pure material. There is also the possibility of detecting interesting species in a provisional rough or enrichment culture (p. 11) which would have been overlooked if we had placed reliance on material from nature only.

Pure cultures are sometimes only obtained after a tedious and wearisome procedure. If material freshly brought from its natural habitat is used, the species required to be raised in pure culture may be lost before success is attained. If we start with a healthy culture, however, successive attempts are possible, and we can learn from failures until our efforts are crowned with success.

At any rate the time spent in growing a species in the laboratory and observing it more closely than would have been possible in the natural habitat is not lost, even if a pure culture is not achieved. Such cultures, especially when originating from single cells, may throw light on certain characteristics, which had not been previously noticed, and may sometimes even prove more helpful than pure cultures.

The following matter deals first with the employment of preparatory mineral mixtures, then with that of special enrichment cultures, and finally with the soil-and-water culture method.

A. *Preparatory cultures in nutritive solutions*. The use of solutions of inorganic salts in preparatory cultures is advisable for various reasons. Their composition can often be made to approximate to that of the natural medium. Such solutions, especially if used in a highly dilute form, have the effect of securing multiplication of species which thrive in pure waters or other localities devoid of organic matter.

If the solutions of Knop, Detmer or Beijerinck are added to a mixture of aquatic or soil algae, species will, after some time, be found growing in abundance which were only present in small numbers and would scarcely have appeared if pure water had been used. The material thus obtained is suitable for further cultural treatment (Lund, 1942, p. 273; John, 1942).

Inorganic solutions favourable to the growth of a species increase the number of individuals, while bacteria and other unwelcome organisms are diminished in comparison to the original mixture. This is advantageous in starting pure cultures, especially when algal cells are hidden between collections of mineral particles, detritus and organic residue, usually called mud. Moreover, they may have been present, as resting stages which might easily be overlooked.

When using this method, it must not be forgotten that this

preparatory treatment tends to result in multiplication of common resistant species which are to be found almost everywhere, while the more specially adapted species, which are often more interesting from a morphological and physiological standpoint, are lost. But when soil, mud or peat are used, the results are better than might be expected from this point of view. If, for instance, the quantity of soil is large as compared with the nutritive salts, their standardizing influence is modified by the effect of the particular kind of soil used and suited to the needs of the algae living in it. A slightly acid reaction of the nutritive solution, for example, will do no harm to species which prefer an alkaline medium, if the pH is shifted in that direction by the reaction of the soil in question.

The addition of inorganic nutritive solutions in preparatory cultures of planktonic species is also helpful and of use in gleaning preliminary information about large sessile species.

B. *Enrichment cultures.* In contrast to that just described this method encourages the multiplication of species adapted to more restricted conditions. Since there are certainly ecological differences between those algal communities which are dependent on special conditions, it should be possible to favour the growth of certain forms in a mixture, while hindering that of others. This can be effected by choosing appropriate media, H'-ion concentrations, temperatures and so on. Actually not much has been done in this respect, and such methods do not in fact work quite as well as might be expected. The explanation lies in the important role played by competition.

The prevalence of certain species under special conditions, among algae as among flowering plants, depends not only on direct adaptation but, even to a greater degree, on competition. A single species by itself can thrive under circumstances under which it could not live when other better adapted species are present. The actual reason why one species of a mixture prevails after a certain time cannot, in most cases, be ascertained by the available methods.

Algae from natural habitats with a high H -ion concentration (e.g. bogs) may be expected to grow only in media of about the same pH. It is seldom possible to produce enrichment cultures of a single species by inoculating it into a medium of low pH, because all species are adapted to nearly the same pH. They must differ in other ecological respects. The influence of H -ion concentration is the clearest among the external factors, but similar statements can be made about others.

As regards Fe-concentration we are far from having achieved the same insight. Certain species are usually found in localities with a high concentration of iron, but for the most part they can grow just as well in media with a much lower iron content. In interpreting these facts, however, it must not be forgotten that we have as yet no chemical means of determining the concentration of available iron in a natural habitat. Part of the iron found by chemical analysis may be ferric, another part ferrous; one part may be in molecular solution, another in complex, insoluble or colloidal compounds. It is unknown what part can be used by algae and what cannot be used (Uspenski, 1925; Pringsheim, 1930, 1936 a). When iron salts are added to an inorganic medium, the greater part is invariably precipitated, whether it be in the ferrous or ferric state, because ferrous compounds are readily oxidized, and ferric ones are insoluble unless the medium is so acid as to make every kind of life impossible. Even in artificial media the effective iron concentration cannot be determined. Each form of iron above mentioned can be changed into another one when the composition of the medium is altered by the metabolism of the algae. Complex compounds are often formed, either with humic or other organic substances. The mode of utilization of these requires further investigation. To sum up, enrichment cultures of iron-loving species should be possible, but the nature and possibility of utilization of the various iron compounds must be better explored before such cultures can be expected to be successful.

Even less is known about the possibility of obtaining by enrichment species of algae adapted to low oxygen pressure, to high concentrations of sulphuretted hydrogen, to very dilute or concentrated solutions, and so on.

Enrichment cultures of non-algal organisms have been obtained and are valuable as a means of studying their ecology, metabolism and morphology and of starting bacteria-free cultures. It should be possible also with algae to achieve much more than has been done up to now in this respect.

C. *Soil-and-water cultures*. Before adequate data concerning algal nutrition had been obtained, the opinion was frequently expressed that most species require the presence of organic compounds. The subsequent successful growth of a great variety of species in inorganic or 'mineral' solutions lent support to the opposite view.

When beginning investigations into the physiology of algae in 1910 I found that many species, which thrive primarily in inorganic media with or without agar, could not be grown in subcultures. The use of media with soil extract (Pringsheim, 1912, 1936a; cf. also p. 40) often brought about a striking improvement, although it did not always induce multiplication. The claim that every chlorophyll-containing organism can be grown in pure mineral solutions cannot therefore be sustained.

In view of these results attempts were made to imitate natural conditions still more closely than by adding soil extract. This led to the soil-and-water culture method, which proved to be surprisingly helpful in cultivating many kinds of small organisms. In nature algae mostly grow in water which is in contact with soil. The use of soil was first suggested by the success of putrefaction cultures.

In order to favour the multiplication of species adapted to extremely rich nutritive media, J. C. Jacobsen (1910) added small quantities of fibrine to soil and water. If the soil contained algae, capable of utilizing substances produced by proteolytic bacteria, the former appeared in quantity in the water above the soil. Other investigators, by pasteurizing the soil, were able to obtain sub-cultures of certain species (Pringsheim, 1921 b, 1936 b; Schreiber, 1925; Strehlow, 1929).

Such cultures, containing organic substances and soil, imitate the conditions in those natural habitats which are usually described as eutrophic. Where the aquatic flora is rich, the bottom mud with its decaying organic residues fertilizes the water. The soil is used as a substitute for mud. Organic material, such as starch or wheat grains, etc., represents organic remains. By varying the putrefying material and the kind of soil, I tried to obtain enrichment cultures of as many diverse species as possible. Substances mainly composed of proteins, such as cheese, gelatine, egg albumen, etc., proved to be very helpful in cultivating Chlamydo-monads *inter alia*, green species appearing in the light, colourless ones in the dark. Starch-containing material, such as grains of cereals or mere starch, proved equally suitable and often gave better results with Euglenineae and other organisms (Pringsheim, 1936 b).

The best kind of soil is ordinary sandy garden loam, not too rich in clay or humus. Plain sand is unsuitable. Sometimes peat is useful, if an acid medium is desired. A small amount of lime, on the other hand, affords an alkaline medium and should always be added to cultures containing starch, since this, by anaerobic fermentation, gives rise to organic acids.

The employment of large vessels is unnecessary with this 'putrefaction-culture' method, ordinary test-tubes being suitable. The following procedure was mostly employed. About three to five grains of wheat or barley, or a similar quantity of starch with a little calcium carbonate, are placed in the tube and covered with soil to a depth of about 3–5 cm. Water is added to within about 4 cm. from the rim of the tube. After plugging the tubes with cotton wool, they are placed in a cold steam-chamber and slowly brought almost to boiling-point, at which temperature they remain for at least 3 hours. Inoculation can take place the next day and should not be deferred for more than a week or two.

The putrefaction-culture method was the device by means of which cultures of *Polytoma*, *Astasia* and *Chilomonas* were first obtained, preparatory to rearing pure cultures (Jacobsen, 1910; Pringsheim, 1921 *b*). The same method has been used to obtain single-cell cultures of many other flagellates, especially Astasiaceae (Pringsheim, 1936 *b*, 1942), and up to the present was the only reliable one for preparing cultures of green Euglenineae (*Euglena*, *Phacus*, *Lepocinclis* and *Trachelomonas*). Such cultures do not, however, prove altogether satisfactory with many green forms, which also fail to grow in mineral solutions, even with soil extract. To avoid the harmful effects of strong putrefaction, which were believed to be responsible for the lack of success, the amount of organic material was reduced. A single grain of wheat or barley often gave a better result. On the other hand, addition of peptone and other soluble and readily split N-compounds, as recommended by Kniep and his pupils (Strehlow, 1929), was often detrimental and did not afford any improvement as compared with insoluble proteins, such as cheese, gelatine and others (Pringsheim, 1936 *b*, p. 49).

When starch only is used and its amount is reduced to a minimum, green algae—Chlorophyceae as well as Heterokontae, Euglenineae and many chlorophyll-containing Flagellata—grow better, even without any further organic material than that contained in garden soil.

A return had thus been made to the simple utilization of soil and water. Even species which grow well in mineral solutions, with or without soil extract, often show a more vigorous, healthy and long-continued development in a medium containing soil. The difference is most striking with filamentous Heterokontae, Ulotrichaceae, Chaetophoraceae, *Vaucheria* and Euglenineae. The reason may be sought in an interchange between the liquid and the colloidal phases of the soil, which retain valuable nutritive substances and deliver them up slowly to the growing algae, so that the medium is not so quickly exhausted as a small quantity of nutritive solution.

Exhaustion of the medium is one of the chief handicaps against obtaining heavy crops of algae in an artificial medium. In natural waters the concentration of nutritive salts is usually very low, but the quantity nevertheless suffices for the development of a vegetation that is macroscopically visible, since concentration takes place at certain places. The same quantity of nutritive compounds, which in nature may be dispersed over a large area, cannot be contained in a test-tube without the realization of a harmful concentration. Fortunately, however, most algae can withstand much higher concentrations than they meet with in their natural surroundings, especially in pure culture, where competition is eliminated. Nevertheless, the amount of nutritive salts that can be concentrated in a restricted volume is sometimes insufficient to allow of a continuous and complete morphological development. This may be the chief reason why the addition of soil has such a pronounced effect. The properties which render soil extract so valuable are found to an even greater degree in soil-water media.

A further advantage may lie, for example, in the capacity of soil colloids to adsorb substances which might otherwise be detrimental. Moreover, rare essential elements may be present in the soil, and finally the supply of iron may be influenced. Humic substances have the property of maintaining iron in solution in the form of complex compounds, thus converting insoluble precipitated compounds into soluble ones. All these features co-operate to produce the favourable effects of soil-and-water cultures, which contribute materially to the facility with which healthy cultures of algae can be grown (Mainx, 1927b; Pringsheim, 1936a).

The soil-and-water culture method can be modified in still other ways than by the addition of putrefiable substances of various kinds. The procedure adopted should always be considered in relation to natural conditions. For example, a reddish clay soil, seemingly poor in humus, proved to be excellent for the cultivation of *Dinobryon sertularia* and other Chrysophyceae inhabiting oligotrophic waters and not growing in any other medium. Earth, rich in humus, provided enough nutritive material to nourish *Desmarella moniliformis* Kent, *Microbotrys pusilla* n.sp., and other holozoic planktonic flagellates which failed to multiply when starch or wheat grains were added. For *Carteria acidophila*, needing very dilute and acid media (Fritsch and John, 1942, p. 374) that are very quickly exhausted if only inorganic salts are present, *Sphagnum* or root peat with distilled water is best. It appears that such peat will provide for months nutrient substances which are adsorbed on its large surface and are progressively set free as they are utilized.

By mixing garden earth with peat or covering peat with soil, the pH can to a certain extent be regulated, while at the same time the algae are provided with humic substances which often have a very beneficial effect. *Sphagnum*, even if sterilized, seems always to have a harmful influence, while well-humified peat from the lower layers of bogs is often very useful. In other cases mixtures of peat with chalk or lime-containing sand and other combinations are helpful, the scope of possible variations being unlimited.

3. OTHER PREPARATORY TREATMENT

A. *Centrifuging*. Algal and flagellate cells are usually denser than the medium in which they are suspended, so that by rapid rotation they can be concentrated at the bottom of a centrifuge tube. Solid reserves, like starch or paramylon, increase the density, while oil has the opposite effect. Bacteria take much longer to settle than most algal cells, owing to their minute size and consequent large surface. If a mixture containing both is centrifuged for a restricted period a partial separation is effected. By repeating the process the number of bacteria relative to that of algae can be reduced (Mainx, 1927 *b*, p. 326; Cleveland, 1928). This treatment may therefore be useful in preparation for plating and increases the likelihood of successful isolation and elimination,of bacteria, but it cannot be repeated sufficiently often to free algal material altogether from accompanying germs. One of its advantages lies in the possibility of concentrating cells from a scantily populated fluid, another in dealing with many cells at the same time.

B. *Tactic movements*. The tactic movements of algal cells have long been used to attain the same end as centrifuging (cf. Bold, 1942, pp. 101–2). Geo- and phototactic movements are useful as a means of obtaining concentrations of algal zoospores and of flagellates, as well as their separation from germs which respond in a different way. By transferring part of such a concentration to a vessel with sterile water a certain measure of purification can be attained, and by repeating the process progressive improvement can be achieved. In this way relatively pure material is obtained which is suitable for inoculating agar plates or liquid media. Chemotactic concentration is of less use, although 'aerotaxis' is sometimes of value. Whenever organisms with flagellar movements are being handled or investigated, care must be taken not to interfere with their geo- and phototactic irritability. These responses may be troublesome or helpful, according to circumstances, but they are always a sign of good health and should be carefully observed.

Tactic movements will be displayed in dishes, watch-glasses or capillary tubes. The aim is to single out cells from a mixture, so that they become separated from the majority of the other organisms present and can then be used to inoculate cultures. The advantages of this method are about the same as with centrifuging. It can, of course, only be applied to tactically sensitive cells which, at the same time, mostly have a clean surface. It is not always easy to put this kind of purification into practice, since most zoospores are motile only during a restricted period. Both they and flagellates tend to come to rest and to become attached to the walls of the containing vessels when transferred to another medium, especially when this procedure is repeated several times.

C. *Isolation of single cells.* The methods mentioned under A and B are less important than formerly, since a third method, the oldest of all and of more general application, has been further developed. This is the so-called pipetting method, which consists in transferring single cells with capillary pipettes into vessels containing a suitable medium.* It is superior to other cleaning methods, because it provides a reliable means of separation of algal species from a mixture,

* Picking up of individual cells under the microscope was probably first carried out by Miquel (1890–2), who in this way obtained cultures of diatoms containing only one species. Pipetting of single cells preparatory to plating is mentioned by Schramm (1914, p. 32).

and affords cultures originating from one individual cell. Furthermore, such isolation may by repetition be made so effective that eventually bacteria are completely absent. The details of the method are given in another chapter (cf. p. 70), but a few points may be mentioned here.

In order to separate algal cells from other organisms, it is best to use fine capillary pipettes (cf. Hildebrand, 1938, pp. 642, 663) with rubber caps. In dealing with simple and branched filaments, thin glass threads or hooks are often more suitable. They are preferable to metal wires, to which algae sometimes adhere so firmly that they cannot, without damage, be introduced into the culture medium. These glass implements are easily made by drawing out pieces of glass rod or tubing and repeating the procedure over a very small flame. To obtain bends the horizontal thread is heated at some distance above such a flame till the part beyond the point of softening drops into a vertical position. By allowing the end to come in contact with the flame for an instant a small knob is formed which will prevent filaments from slipping off or injury to the cells by sharp edges. Vischer (1937, p. 191) recommends the drawing of filaments over a sterile agar surface in order to cleanse them prior to the formation of colonies.

Mechanical isolation under the microscope, usually achieved by pipetting, has some striking advantages compared with other methods, some of which may be mentioned.

Several or even all the species in a mixture can be separated and inoculated into different vessels with different media. Recognizably healthy cells can be used for starting cultures.* Fungi, amoebae and unwanted algae, such as Diatoms and blue-green forms, which often prove very trouble-

* Bold (1942, p. 99) does not altogether recognize the most valuable feature of the method, which is to isolate certain algae at will; no other method provides for this.

some by spreading over the agar surface or suppressing the growth of desired species in liquid media, can easily be excluded.

The ratio between algal and bacterial cells can also be made much more favourable than it was in the original mixture. The cells can immediately be placed under suitable conditions and isolation is quickly performed. No extensive dilution series, with sets of many flasks (Kufferath, 1930, p. 118), are necessary for singling out individual cells. The flasks are much better employed in testing out various media.

In applying the pipetting method the living material is watched under a binocular microscope, during which procedure some familiarity is made with it which will prove helpful. The older methods of isolation by dilution or spreading do not allow of close observation and are much more dependent on chance.

There are also some disadvantages. It is necessary to acquire some skill in handling the implements, and limits are set by the size and certain other properties of algal cells. The efficiency of the pipetting method is most clearly revealed in handling motile flagellates or zoospores. These are not only for the most part free of bacteria, but they catch the eye and are therefore more readily recognized under the binocular microscope than motionless cells. If the latter be small like those of *Chlorella*, other methods are advisable. Flagellates of similar dimensions (e.g. *Cyathomonas* or *Rhabdomonas*) or minute zoospores are easily picked up. Cells which adhere to the surface of the glass, like those of many Cyanophyceae, Diatoms, Desmids and Euglenineae, offer an appreciable resistance when being sucked into the capillary tube. But all these difficulties can be overcome with patience, and the final results will repay the labour.

The pipetting method is in no way restricted to free-swimming cells. Many non-motile species, e.g. planktonic and bottom-living forms, can hardly be isolated by any other method. They can, however, be picked up with a capillary tube, even if there are only very few of them. Afterwards they can either be transferred to an agar plate or first subjected to preparatory culturing. In the latter case many individuals of identical origin are raised and become available for making pure cultures.

D. *Methods of obtaining clean cells.* All efforts to separate algae from other micro-organisms, especially bacteria, would be futile did they not possess stages, the surfaces of which are clean and free from epiphytes. Many filamentous algae, as well as others, are usually covered with bacteria. *Vaucheria*, *Cladophora*, *Enteromorpha* and all Phaeophyceae and Rhodophyceae hitherto examined were found to be infected in this way, although most Conjugales, such as *Spirogyra*, *Zygnema* and Desmids, appear to be free from them, if in a state of healthy growth. These can therefore be isolated in pure culture, if inoculation of sterile media is undertaken directly in the natural habitat (Czurda, 1926). Transportation in closed vessels, however, induces an enormous multiplication of bacteria, which cannot be removed by washing. Schramm (1914, p. 35) started pure cultures of *Protosiphon*, with the help of the cysts, which were liberated by rupturing the cell membrane. In *Eremosphaera* also, pure cultures have been obtained by freeing young daughter cells from the mother cell wall. They proved to be free of adhering germs (Mainx, 1927*a*).

Among Chlorococcales, such as *Chlorella*, *Scenedesmus*, *Oocystis*, etc., occasional individuals out of a large number may be devoid of bacteria as a result of such frequent division that the cells, when freed from the parent membrane, are not infected. Such cells give rise on agar to colonies which can be used to obtain pure cultures.

For the most part moving flagellates or zoospores can be relied upon to be clean in this sense. It is therefore essential to know how to obtain stages which swim with the help of flagella (cf. Schramm, 1914, p. 30). True algae only occasionally form zoospores, but flagellates also are often found multiplying in a non-motile state as so-called palmelloid stages, in which the cells are surrounded by mucilage envelopes of varying thickness. 'The conditions calling forth the development of these palmelloid stages in nature are hardly known' (Fritsch, 1935, p. 16). Cell division may continue for a long time in this state. Since the formation of envelopes and escape from them are opposite occurrences, it is not to be wondered at that the conditions determining these events are also insufficiently known.

Not much more is to be found in the literature regarding the conditions which call forth the formation of motile stages in true algae. The results of certain investigations were published as early as 1896 by Klebs, although not all of these have proved capable of repetition. The nature of the reaction depends on the physiological condition of the living material. In general, however, Klebs's results have been confirmed by Gerneck (1907), Freund (1908) and others.

It is usually to be expected that zoospores, like motile stages of flagellates, will be formed

(1) when algae are transferred from moist to liquid media, or

(2) from flowing, well-aerated, water to a vessel in the laboratory;

(3) when algae are provided with fresh nutriment after the old medium has been exhausted;

(4) when algal material is kept in the dark for more than a single night and subsequently exposed to a suitable illumination.

Freund (1908) found that in *Oedogonium pluviale*, as in many other algae (*Vaucheria repens*, *Hydrodictyon reticulatum*, *Protococcus viridis*, etc.), the formation of zoospores is readily induced by transferring material grown in inorganic nutrient solutions to distilled water or vice versa.

As a rule, changes in the surrounding conditions tend to promote swarmer formation. In some species this can be induced in a short time, in others only after a longer period. The former is true, for example, of *Oedogonium* and Ulotrichaceae, the latter of *Vaucheria* and *Hydrodictyon*. Sometimes changes other than those above mentioned prove effective, for instance, a rise in temperature, dilution of the medium, aeration, etc.

Flagellates forming palmelloid stages, which are often of some degree of permanency, are found in all classes, e.g. in Chrysophyceae such as *Chromulina*, in Cryptomonadineae such as *Cryptomonas*, and in Euglenineae such as certain species of *Euglena* (e.g. *E. sociabilis*, *E. pisciformis*, *E. viridis*). Among Volvocales, *Chlamydomonas intermedia* and *Haematococcus pluvialis* tend to live in the palmelloid stage. In *Haematococcus* (Pringsheim, 1914 *b*, pp. 428 et seq.) this is determined by exhaustion of the nutriment. Free movement ceases in the absence of a single indispensable element, which had previously been present in the relatively smallest quantity, this being the 'limiting factor' for multiplication. Addition of a suitable compound containing this element is sufficient to start movement again. It may well be that these experiments afford an example of what is a general rule (Freund, 1908; Pringsheim, 1913, p. 46; Glade, 1914, p. 333).

In most instances the reasons for the cessation and commencement of swarming are still unknown. It sometimes suffices to change the position of the culture vessels to induce solution of the gelatinous envelopes of the palmelloid stages, and this may happen within an hour. In other cases cells squeezed out of their envelope by the cover-slip immediately start swarming. By transferring material of the flagellate from agar cultures to a dilute iodine solution, it can be decided whether flagella are already present or whether they are formed only after transference to a liquid medium. In

Polytoma and *Chlamydomonas* they are usually preformed, in *Haematococcus* and in *Euglena* they appear to be absent. The time that elapses until swarming sets in is therefore much shorter in the former than in the latter. Many more data require to be collected to provide a guide as to the ability of algae and flagellates to form free-moving stages, so that cells devoid of bacteria may be obtainable with a view to the preparation of pure cultures.

It is hardly likely that algae, possessed of the faculty of forming zoospores, will fail to do so in a series of subcultures in various media. Zoospores betray their presence as a green line near the surface of the liquid on the side facing the light; in test-tubes they form a vertical line, where the light is concentrated by the refraction of the water which operates like a lens. After a certain, often rather short, interval the zoospores adhere to the glass and germinate. The period during which isolation could be performed is then terminated, although the existence of phototactic zoospores is disclosed. A repetition of the experiment, combined with careful observation, will afford the opportunity of securing a pure culture of the species concerned.

In Diatoms, which do not form flagellar stages, we depend on the occurrence of rapidly moving cells, and in Cyanophyceae on hormogonia, for obtaining pure cultures. Such structures creep over the surface of agar plates or through the actual jelly, and thus in rare instances free themselves from adhering bacteria. With some luck such stages may be found. Species not forming any motile stages are hard to free of bacteria. Pure cultures of motionless species of Cyanophyceae have not to my knowledge been achieved, although many immotile Ulotrichaceae and Chlorococcales have been grown in pure cultures.*

* Concerning the use of chemical compounds and of ultra-violet light to kill contaminating organisms, cf. p. 95.

CHAPTER III

IMPLEMENTS AND MEDIA FOR
GROWING ALGAE

1. WATER AND CONTAINING VESSELS

A. *Water*. One of the first things to consider in the
cultivation of algae is the water. Up to 1912 tap and spring
water were usually used, because distilled water proved to
be fatal. Although Naegeli as early as 1893 indicated that
traces of copper might be the cause of its harmful quality,
resulting from the use of a metal distilling apparatus, the
method of redistilling water by condensation through a glass
tube was introduced much later* (Pringsheim, 1912, p. 327).
Molisch (1895) and Benecke (1898) had used a platinum
cooling tube to obtain salt-free water, but not for removing
heavy metals. Molisch's apparatus at least was not efficient
in this respect, owing to the shortness of the platinum tube,
as I know from personal use of it; by distilling non-toxic
water through it satisfactory results could, however, be
obtained.

Glass is a cheaper material to employ than platinum;
it is best to use Pyrex or Jena glass, but, if still greater care
is necessary, quartz is now available. In most cases, and
especially if silicon is to be avoided, pure tin is equally good
and not so fragile. I used it for a long time (cf. also Hutner,
1936–7, p. 94). The water distilled from it should, however,
be tested for non-toxicity, which can suitably be done with
Spirogyra (Naegeli, 1893). Chlorine, which is used for dis-
infecting tap water in many water supplies, must also be

* The roots of higher plants do not appear to be as liable to poisoning
by copper as are many algae.

guarded against. In Prague, and even more so in Cambridge, tap water was found to be definitely poisonous for this reason, but not the distilled water prepared from it. I doubt whether the same would be true of soft waters.

When distilled water is redistilled from a glass flask, boiling is often retarded, and sudden ebullitions alternate with periods of quiescence. To overcome this difficulty it is sufficient to put a small piece of pumice into the water.

It must not be assumed without tests that nothing but steam comes over when water is distilled. Volatile substances from the air (ammonia and its organic derivates, fatty acids and others), which have been taken up by the water, often pass over with it. In order to destroy such compounds potassium permanganate and sulphuric acid are added to the water. More are added when the solution turns brown. The presence of volatile substances in the distilled water would introduce errors into experiments on nitrogen-fixation and with autotrophic organisms, and in other kinds of research on nutrition.

The possibility must not be overlooked that traces of non-volatile dissolved substances may pass over with tiny water droplets which are dispersed as the steam bubbles burst. This can be tested by adding a coloured dye to the water. It has been suggested that droplets can be prevented from reaching the receiving vessel and caused to flow back, by allowing the steam to pass through a glass tube containing short pieces of hard glass tubing before entering the condensation tube. When testing this device, I found no difference in the biological effect of the water as compared with that distilled with a simple tube, the first part of which was bent up and down in the shape of a siphon.

Sometimes water, other than distilled water, is useful. When very large amounts of nutrient solutions are required, distilled water is too expensive and can often be replaced by tap or spring water. If the latter is hard, it should first be boiled or treated in a permutit softening apparatus. Chlorine can likewise be removed in this way or by using a filter of charcoal. Sometimes the calcium bicarbonate in hard water is useful, especially in cultivating algae growing in natural habitats rich in lime or chalk, or for neutralizing the acid substances contained in a nutrient solution.

Special consideration must be given to sea and brackish water. No artificial mixture has hitherto proved to be the equal of natural sea water for biological purposes. If, however, this is not available or is ruled out by the special needs of the research, recipes given in the literature (Küster, 1913, pp. 14 et seq.) can be used instead. The most convenient mixture that has often proved suitable, but is only approximately similar to natural sea water, is the following: NaCl 3 %; $MgCl_2$ 0·4 %; KCl 1 %; $MgSO_4.7H_2O$ 0·5 %; $CaSO_4$ 0·1 %. For the culture of algae this solution may be 'fertilized' with 0·01 % $NaNO_3$ and 0·001 % K_2HPO_4. I use hard tap water in its preparation. If the water is soft, or distilled water is used, a small quantity of $CaCO_3$ should be added after complete solution of the ingredients. This artificial sea water is much improved by the addition of soil extract (cf. p. 40). A similar solution, prepared with natural sea water, was described by Føyn (1934, p. 7; cf. also Schreiber, 1928, 1931; Haemmerling, 1931, 1934) as 'Erd-Schreiber' solution and has been used in the cultivation of marine algae. But, if employed with soil extract as recommended, artificial sea water is nearly as suitable as the natural fluid. Allen (1914, p. 417) found that the Diatom used in his work failed to grow in artificial sea water, but did so when as little as 1 % natural sea water was added; with 4 % of sea water multiplication was even better than in natural sea water alone.

B. *Culture vessels*. Culture vessels made of glass are nearly always employed. Küster (1913, pp. 8 et seq.) deals in detail with the properties of the different kinds of glass, and there is no necessity to repeat or revise his statements, because hard glass of low solubility is now purchasable everywhere. Further details will be found in Ondraček's (1935) and Bold's (1942, p. 75) papers. The best-known makes are Pyrex and Jena glass, which, although more expensive than ordinary glass, last longer and are therefore on the whole cheaper.

For the majority of algal investigations large vessels are not requisite and test-tubes suffice. They are cheap, easily manipulated and do not take up much room. Thick-walled test-tubes of the ordinary size (⅜ by 6 in. or 16 by 160 mm.), with rims, can be used for research and routine work alike.

Ordinary test-tubes are sometimes useful. Small, cheap and light tubes, which can be stored compactly, are preferable when it is a question of maintaining a collection of established strains. When sent by post they are light and they can be sealed, which is the best way of preventing contamination of the cultures in transit. Moreover, a smaller quantity of medium is required. In rare instances substances released from the glass are helpful. Thus, silica for diatoms and other algae is most simply provided by the use of soft glass. Sometimes other, not yet fully known, substances are provided by the glass (Czurda, 1933).

If species of larger dimensions are to be reared or an appreciable quantity is required, or if, again, the ratio of surface to volume of the fluid is to be greater, Erlenmeyer flasks, jam jars, milk bottles and the like can be used. These should be covered with well-fitting lids, or at least with tightly fastened paper caps, if plugs of cotton-wool cannot be inserted. Flasks and jars of Pyrex and Jena glass used for preserving fruit are obtainable and generally preferable to ordinary glassware.

All glass contains various silicates, and none of these is completely insoluble. If the mineral requirements are being investigated, and especially in the case of silicon, the use of glass vessels is open to suspicion. For such purposes the simplest device is still that suggested by Molisch (Richter, 1906, pp. 27, 37; Küster, 1913, p. 11), i.e. ordinary glass vessels, covered inside with a layer of pure paraffin wax. The vessels, usually Erlenmeyer flasks, are first dried in an oven. Then the source of heat is turned off and small pieces of hard paraffin are inserted and allowed to melt. After that the flasks are rotated under a running tap until a thin layer of wax covers the entire inner surface which will be in contact with the medium. The vessels are sterile and should be covered with a sterile glass cap till wanted.

To ensure the absence of silicon and other elements present in the glass, platinum- or tin-distilled water has to be used for experiments with paraffin-covered vessels. The medium can be prepared, as a concentrated stock solution, in a platinum cup and sterilized in the same vessel. Part of the solution can be transferred with a pipette of hard glass to the paraffined flask and diluted with newly distilled water, which is of course likewise sterile. At room temperature the solution of substances from glass, especially if this has been in use for some time, is much slower than at 100° C., so that it is not very material if the medium comes in contact with such glass for a short time.

If investigations as to the chemical elements requisite for the growth of an alga are being undertaken, for which it is inadvisable to use hard glass, and the paraffin method cannot be followed, resource is had to quartz glass. This is manufactured in every shape and size and is no longer excessively expensive. Unless broken or heated in the presence of strong alkali, it lasts indefinitely and can be used in the same ways as hard glass. Coarse porous plates of porcelain and other kinds of earthenware are useful for growing rock algae.

Rubber stoppers and tubing are not inert, and only those of the best quality should be employed; even these contain sulphur compounds which cause trouble. The harmful or beneficial effects of rubber manifest themselves not only when in contact with the culture medium; it also gives off volatile substances. Rubber tubing is not completely impermeable, and coal gas, carbon dioxide, ether- and chloroform-vapour, etc., pass through it in quantities sufficient to influence vitality, growth and reproduction.

Adhesives and plastics, such as sealing wax, picein, paraffin and plasticine, may have a certain influence and should be tested by putting bits into cultures before using them for experimental purposes. The same is true of cement. As paraffin is almost insoluble, any other substance can be protected with a covering of it; for this purpose hard paraffin is preferable, because it is less volatile. Xylol, benzol and other solvents which are poisonous must be avoided.

Metal should be used as little as possible. Platinum, gold and tin are almost without influence, but not so silver. Gold is usually alloyed with silver and copper, tin with lead, which are all poisonous.

At present no other materials come into consideration in connection with algal cultures, but translucent plastics may in future be of service in biological investigations, e.g. as materials for collecting vessels.

2. MINERAL SOLUTIONS

Pigmented algae do not, as a rule, need organic compounds. A few species of *Euglena*, it is true, are stated to require organic nitrogen compounds in addition to the products of CO_2 assimilation, but this is not fully established (Hall, 1939, pp. 3 et seq.). There is, however, no doubt that a great many species of chlorophyll-containing algae thrive much better when provided with organic substances, such as peptone, sugar, fatty acids, etc., than in a medium composed of inorganic salts and water only.

There are many reasons why inorganic or 'mineral' solutions are nevertheless employed:

(1) They provide basic media to which organic compounds can be added according to requirements.

(2) Algae intermingled with bacteria, as they always are in nature, cannot generally be cultivated in solutions containing organic substances, because the bacteria would multiply to such an extent that the algae would be injured.

(3) For some algae organic compounds are not beneficial and may even be detrimental.

(4) Species growing in waters almost completely devoid of organic substances contrast in their nutritional requirements with those found in polluted waters. It is of interest to know how the respective groups behave in organic solutions.

A. *Concentration and composition.* The first nutrient solutions used in the study of algae were those devised by Sachs, Knop, Pfeffer, etc., for the growth of flowering plants in 'water cultures', but for most algae these are too acid. Hence the warnings of Molisch (1896) and others to grow algae in slightly alkaline solutions. The importance of H -ion concentration was then not known. Algae living in acid bog and peat waters of course prefer a pH below 7 and even as low as 3·5, but the nutrient solutions mentioned above are also unsuitable for them, since they are too concentrated for organisms from such habitats.

Apart from the effect of the pH and the concentration of the medium, little is known as to the properties of mineral solutions that are of special consequence to growing algae (Vischer, 1920). The precision advocated in many recipes is apt to be misleading. When various nutrient solutions are compared, one wonders what may be the purpose of the diverse ingredients and of the precise quantities indicated.

Most algae are not readily affected by minute changes in the composition of the medium, otherwise they could not live under natural conditions. Changes effected by the algae themselves are often more decisive than differences between various media, as recommended by various authors and enumerated by Kufferath (1930, pp. 41 et seq.) and others. It is impossible to avoid the conclusion that many of the small differences in the composition of various media are inessential, and Bold (1942, p. 77) rightly refrains from discussing them. The differences concern concentration, nitrogen supply, calcium, iron, sodium, chloride, etc. It is evident that most of these media are only of historical interest, although not all of them need be forthwith abandoned.

The writer has always endeavoured to simplify media as much as possible, with the object of saving time in their preparation, as well as of rendering more intelligible the conditions obtaining in them. Difficulties can often be overcome by experimenting with various concentrations. In dealing with sensitive forms, e.g. desmids, NH_4MgPO_4 was included for the first time in the medium (Pringsheim, 1912, p. 328), whereby a low concentration of three essential elements is secured, while a continuous supply is maintained by a bottom deposit of this almost insoluble salt. In this way the first successful cultures of desmids were prepared.[*] In the cultures of *Micrasterias* (1930, p. 12) another principle was employed (see p. 37 dealing with N-sources).

Recently Chu (1942), in a study of the conditions of existence of planktonic algae, has shown that the majority of such organisms will grow in very dilute inorganic solutions. It is evident that such solutions should be even more dilute than those suitable for Conjugales (Czurda 1926). Chu's media contain only 0·002–0·004 % of nitrate or ammonium salts. His investigations indicate that the concentra-

---

[*] This new principle became the starting-point for a long series of experiments, leading to the use of biphasic culture media, mostly composed of soil and a watery medium (cp. p. 16).

tions of the different nutrient salts must be accurately adapted to the specific needs, if optimal growth is to be obtained, but unfortunately no pH measurements are given in his paper. Altogether seventeen media are enumerated, the first of which is similar to one used by Beijerinck (1898) although more dilute, the concentration of NH_4NO_3 being only 0·0025 %. Starting from this medium, others were developed by the introduction of certain modifications. Medium No. 5 and onwards afforded healthy growth of such plankton algae as *Pediastrum Boryanum, Staurastrum paradoxum, Nitzschia acicularis, Nitzschia palea, Fragilaria crotonensis, Asterionella gracillima, Tabellaria fenestrata* and *T. flocculosa*. All successful media contain carbonate ($CaCO_3$ or Na_2CO_3) or silicate (K_2SiO_3) or both. These salts will neutralize the otherwise acid solutions, so that the pH is adjusted to the degree suitable for planktonic algae. The only exception is constituted by *Botryococcus Braunii*, which grows in medium No. 2 containing 0·0025 % KNO_3 instead of NH_4NO_3 and no carbonate; this should be slightly acid. *Botryococcus* thrives also in No. 13, which differs from No. 2 in some of the ingredients being present in higher concentrations. It cannot be concluded that *Botryococcus* prefers an acid medium because it, like most of the other species, grew best in No. 10 (Chu, 1942, p. 298), the composition of which is as follows:

$Ca(NO_3)_2$	0·004 %
K_2HPO_4	0·001 or 0·0005 %
$MgSO_4.7H_2O$	0·0025 %
Na_2CO_3	0·002 %
Na_2SiO_3	0·0025 %
$FeCl_3$	0·00008 %

The results leave the question open as to how far the beneficial results of the relatively high phosphate concentration in Chu's media are due to its buffering effect. It seems, however, that certain planktonic algae are actually more dependent on a definite mixture of inorganic salts than is usually supposed, so that in future more attention must be paid to the absolute and relative concentrations of ingredients. Chu's studies show the value of varying inorganic media when it is a question of growing algae which fail to multiply in the mineral solutions ordinarily used.

A few nutritive solutions, which have often proved suitable, are given below:

(1) Knop, modified

KNO_3	0·1 %
$Ca(NO_3)_2$	0·01 %
K_2HPO_4	0·02 %
$MgSO_4.7H_2O$	0·01 %
$FeCl_3$	0·0001 %

(2) Molisch

$(NH_4)_2HPO_4$	0·08 %
K_2HPO_4	0·04 %
$MgSO_4.7H_2O$	0·04 %
$CaSO_4$	0·04 %
$FeSO_4.7H_2O$	1 drop of 1 % solution to 100 c.c.

(3) Pringsheim (for *Micrasterias*)

KNO_3	0·02 %
$(NH_4)_2HPO_4$	0·002 %
$MgSO_4.7H_2O$	0·001 %
$CaCl_2.6H_2O$	0·00005 %
$FeCl_3$	0·00005 %

(4) Beijerinck

NH_4NO_3	0·1 %
K_2HPO_4	0·02 %
$MgSO_4.7H_2O$	0·01 %
$FeCl_3$	0·0001 %

All these solutions can be used in dilutions of 1/2, 1/4, 1/10.

B. *The nature of the nitrogen supply*. One of the chief respects in which mineral solutions vary is the form in which nitrogen is supplied. Nitrites need not be considered, because they are of little value or even poisonous, so that only nitrates and ammonium salts remain.

Certain early experiments (Pringsheim, 1912) gave the impression that some algae preferred ammonia, others nitrates, a result which proved to be due to secondary influences. Later it was concluded (Pringsheim, 1926 *a*, p. 298) that no chlorophyll-containing organism had so far been found which could not utilize either type of inorganic nitrogen, and this is still not in conflict with any more recent experience.

Nitrates appear always to be reduced to ammonia before they are used for synthesizing amino-acids, from which proteins are formed. We can therefore not expect to discover organisms capable of using nitrates only, although the opposite cannot be ruled out. For colourless forms a supply of ammonium compounds is often essential, if no organic N is available; this is so with *Polytomella*, *Polytoma*, *Hyalogonium* and *Astasia longa*, but no green form is known to have such requirements. *Chlorogonium* can use nitrates, not only in the light, but also in the dark (Pringsheim, 1934 *b*, p. 148; 1937, p. 643), a fact which disproves the hypothesis that nitrate reduction must be connected with photosynthesis (Molliard, 1925, p. 120; Kostytschew, 1926, p. 149). Other green forms, capable of multiplying in the dark (e.g. *Euglena gracilis* and species of *Chlamydomonas*), appear to utilize nitrates only in the light, and in the dark require to be supplied with ammonia or organic N-compounds.

When both ammonia and nitrate are present in the form of NH_4NO_3 (Beijerinck's and Benecke's solutions), ammonia only is usually absorbed, or it is taken up preferentially so that there is an accumulation of acid and the pH falls. The use of ammonium nitrate, therefore, affords no guarantee that the pH will not alter. It seems, however (Trelease and Trelease, 1935), that there are differences in this respect. Thus, rice is supposed to require most of its nitrogen in the form of NH_4 (Trelease and Paulino, 1920), while wheat uses both nitrate and ammonium simultaneously so long as the reaction remains definitely acid. Similar differences may exist among algae, although this may be true only of acid solutions. As a rule higher plants show a preference for ammonia in alkaline and for nitrates in acid solutions.

In order to prevent too great a decrease of the pH when ammonium salts are provided, a surplus of calcium carbonate is often very useful. In equilibrium with a liquid medium this gives an alkaline reaction of about pH 8. In test-tubes with the usual quantity of liquid which is not stirred, an acid reaction may appear nevertheless near the top of the fluid, where absorption of CO_2 from the air and utilization of ammonia lower the pH. On the other hand, uptake of CO_2 during assimilation—even from carbonates—in the lower layers will cause the pH to rise markedly (Allison, Hoover and Morris, 1937), especially if the organisms present are living near the bottom deposit of $CaCO_3$.

The only method of keeping organisms under constant conditions is to cultivate them in a flowing medium (Waren (Waris), 1933, 1936, 1939). When this is impossible, it should be remembered that large and broad vessels admit of a more prolonged growth than small and narrow ones like test-tubes. These latter are, however, so handy that they are nevertheless for the most part preferable and, even in them, certain species remain alive for a year or more, if freed from competition with other algae.

Another way of preventing excessive pH changes is to supply nitrogen in two different salts, preferably KNO_3 and $(NH_4)_2HPO_4$, the concentration of the latter being about one-tenth of the former (Pringsheim, 1930, p. 12). Since ammonia is first used, the reaction will be slightly more acid when this is exhausted. This increased acidity will, however, be checked because, as the ammonium concentration is reduced, nitrate-N is taken up and this tends to shift the pH in the opposite direction. Media with this combination have proved useful in rearing delicate algae, which lose their fresh colour in solutions containing nitrate or ammonium salts only, as well as in those containing ammonium nitrate. Trelease and Trelease (1935) have shown that wheat may consume ammonia and nitrate at about the same rate, so that NH_4NO_3 is completely used up and no change of pH ensues; this takes place, however, at a low pH (4·3–5·1), which would be harmful to most algae. There is no evidence that a similar result can be obtained with any algal species.

C. *Other chemical elements*. Among the chemical elements contained in many nutritive solutions, chlorine is certainly unnecessary. The role of calcium is not clear. Many species of algae multiply abundantly in solutions of pure salts among which Ca is excluded (Molisch, 1895; Pringsheim, 1926*b*; Warén, 1926, 1933, 1936). On the other hand, it may not be taken for granted that any species will grow if Ca be completely absent or even merely reduced to the extent possible with elaborate modern methods.

If a growth of algae is the sole aim, Ca should always be provided. It should mostly be added in a very low concentration, because it tends to precipitate phosphates and iron unless the solution be definitely acid, which would be harmful to most species. Unfortunately, attempts to prevent a change of pH and to hinder the precipitation of indispensable chemical elements often conflict with one another, and it is not always easy to prepare a really satisfactory medium.

Phosphates, which are so useful for buffering H-ion concentration, are mostly not very effective when present in amounts which will not hinder multiplication. In nature bi-carbonates frequently act as buffers, but they are difficult to introduce into experimental conditions. Acetates are also useful as buffers, but are liable to be used up in the metabolism of the algae. Other buffers employed in physico-chemical experiments are poisonous.

Even when the six indispensable elements K, Ca, Mg,

Fe, S and P are provided as suitable salts in the right concentrations and at a suitable pH many, and possibly all, algae fail to grow in successive subcultures or only exhibit a very limited growth. The explanation may lie in the necessity for the presence of certain other elements in extremely low concentrations. Several facts point in this direction.

In Czurda's experiments with *Spirogyra* and other Conjugales (1926), vessels of insoluble material, such as Jena glass, afforded much poorer growth than those made of ordinary glass, and did not admit of repeated subculturing. The difference cannot be explained only by the difference of pH, which is increased by the alkaline substances in ordinary glass. It must be concluded that special elements indispensable for the growth of Conjugatae were lacking in Czurda's solution, but were provided by the material of ordinary glass.

Certain species of algae, for instance pronounced planktonic forms, will grow only in solutions of a more complex composition than those of Knop, Beijerinck, etc. This may be due partly to the necessity of buffering or balancing the effect of different ions, but it seems that addition of supplementary elements also plays a role. Elements, which are supplementary to those usually present in nutritive solutions and which may be supposed to be required by all or some algal species, can either be added at random or provided in a definite way. If ashes of organic tissues are added, we can almost certainly rely on the presence of every element necessary for organic life. By dissolving such ashes in dilute hydrochloric acid we obtain solutions suitable for use after neutralization.

The alternative is to prepare solutions of salts of the elements suspected of being indispensable. These are iron, copper, zinc, manganese, molybdenum, iodine, silicon, etc., all of which have been found effective in influencing the growth of higher plants and fungi. A preparation known as

A-Z solution,* after Hoagland, has been used successfully by Schropp and Scharrer (1933), for example, with water cultures of higher plants (cf. also White, 1938). Very little is known as to the requirements of algae with respect to the elements contained in this solution.

Soil and peat extract, which often have a remarkably favourable influence on micro-organisms, especially when mixed with mineral solutions (see the next section), are likely to contain such trace elements.

It is not suggested that the hints given above cover the whole field of the preparation of suitable mineral solutions. On the contrary they represent scarcely more than the rudiments. The fact that much more is known concerning nutrient solutions for higher plants shows that there is still a wide and promising field for investigation with respect to algae.

3. EXTRACTS OF SOIL, PEAT, ETC.

Extracts containing humus substances, which are widely used in culturing both marine and fresh-water algae, were first introduced by the writer about thirty years ago. They have proved very helpful, in fact many species cannot be grown in their absence (Pringsheim, 1912, p. 326).

The fact that many algae fail to multiply satisfactorily in mineral solutions raises the question why they thrive so much better in their natural habitats. The search for substances

* This solution contains:

LiCl	0·5	$MnCl_2.4H_2O$	7·0
$CuSO_4.5H_2O$	1·0	$NiSO_4.6H_2O$	1·0
$ZnSO_4$	1·0	$Co(NO_3)_2.6H_2O$	1·0
H_3BO_4	11·0	TiO_2	1·0
$Al_2(SO_4)_3$	1·0	KI	0·5
$SnCl_3.1H_2O$	0·5	KBr	0·5

Grams in 18 litres of distilled water. 1 c.c. of the solution is added to a litre of nutrient solution. It may be doubted whether the salts are present in a concentration approaching the optimum.

that would promote growth seemed most likely to be solved by ecological considerations. Most organic substances could be ignored, since they are either lacking in the natural habitats or, if present, would be quickly destroyed by other organisms; nor can they be used in cultures not freed from bacteria.

Humus compounds, however, seemed to offer greater promise, since they occur in most natural habitats, though only in low concentrations. From the first the addition of these substances proved very advantageous, if prepared in the right way (Pringsheim, 1913, p. 40; 1926 a, p. 302; 1936 a). The extraction of soil or peat can be effected with hot or cold water. In the latter case the extracts contain very little humus substances and are comparable to dilute solutions prepared by heating.

The humus cannot be extracted with alkaline solutions, because, as a result of neutralization, colloidal humic acids are precipitated. But, by treating soil with boiling water, a brown liquid is obtained which is effective. Any garden soil is suitable, but old leaf mould that has undergone decay for at least three years is best; if there has been less decay, the growth of many bacteria is encouraged. If possible the earth should contain no clay, the presence of which makes it difficult to obtain a clear solution. Acid soil can be neutralized by the addition of chalk before boiling.

The quantity of soil and of water used for hot extraction, and the amount of extract to be added to an algal medium, depend on the amount of humus dissolved during heating. With leaf mould about twice the volume of water is added, and for use the extract is diluted to between 1/10 and 1/50.

Soil and tap water are heated in a steam chamber for an hour or in an autoclave at 1 atmosphere above atmospheric pressure for a few minutes. After cooling, the brown solution is poured into a glass vessel. Floating particles are removed with a sieve or by any other method of straining. Filtering is wearisome, and it is much more convenient to allow the turbid particles to settle. This takes some time and, as putrefaction might set in, an antiseptic such as ether, chloroform or toluol should be added, and the bottle firmly stoppered. Shaking with cellulose wadding will also remove turbidity, although it is doubtful whether it is always effective.

The quantity of clear extract needed is taken up with a pipette, care being taken not to include the antiseptic which, in the case of ether and toluol, is at the surface, and in the case of chloroform, is at the bottom of the vessel. The small amounts of these substances dissolved in the medium evaporate during sterilization. If the antiseptics cannot be completely separated from the extract, they can be filtered off through a previously soaked filter-paper. All soil extracts should be carefully autoclaved, because soil often contains heat-resisting bacterial spores which are not killed by the previous heating or by the antiseptics.

Every soil extract should be tested before use. The pH of a solution equal to that to be introduced into the medium is determined and a number of cultural experiments are undertaken.

It is not fully known what properties are responsible for the beneficial influence of soil extract. It has a considerable pH buffering effect and prevents some substances, especially ferric compounds, from being readily precipitated. Its usefulness may in part be due to these properties (Pringsheim, 1936 *a*, p. 117). Although soil extract is not quite as effective as soil itself, its influence is similar, so that it is very useful wherever a purely liquid medium is required. It can sometimes be replaced by caramel (Pringsheim, 1936 *a*, p. 115) or by iron compounds added after sterilization (Provasoli, 1937–8, p. 27).

If a low pH is required, as for organisms from bogs and swamps, peat extract can be used instead of soil extract (Wettstein, 1921).

Some organisms (*Carteria acidicola* Fritsch and John, *Euglena mutabilis* Schmitz) exhibit much better growth if the peat particles are not removed, but left at the bottom of the test-tube. This constitutes a parallel to the superiority of soil-and-water cultures over all liquid media, even those with soil extract. For the most part, however, peat particles are harmful, also to organisms adapted to a low pH.

In the preparation of peat extract only old peat must be employed, *Sphagnum* and root peat being almost equally suitable. Fresh *Sphagnum* is poisonous. The extract is made with glass-distilled water. It contains few dissolved substances and has a pH between 3·5 and 5. Higher values can be obtained by mixing peat extract with soil extract.

Extracts prepared from organic substances are dealt with in the next section. Those discussed above can be regarded as leading over to them because they contain only very small quantities of putrefying substances. This is true also of extracts of moss, straw or dead grass which sometimes prove favourable.

4. ORGANIC MEDIA

A. *Organic compounds.* It is the province of research into the physiology of nutrition, rather than pertaining to the methodology of pure cultures, to determine the organic compounds that are suitable for the growth of a given alga. Their utilization, however, requires discussion here because many algae fail to grow properly without organic substances, and pure cultures can scarcely be obtained without abundant multiplication. This is specially relevant, since pure cultures mostly originate from single cells, which do not seem to give rise to a dense culture as readily as a large quantity of inoculated algal material. Moreover, the simplest way of detecting bacteria, and thus of making sure that a culture actually is and remains pure, is to inoculate it into a medium containing organic substances. It is therefore advisable to employ such media also for stock cultures of species which do not require organic substances. For saprophytic species lacking chlorophyll the importance of organic substances is obvious.

From a theoretical standpoint numerous compounds might be tested, although very poisonous ones can be rejected at the outset. In investigating a similar problem, that of chemotactic attraction, many compounds were found effective that would not have been suspected of being so (Pringsheim and Mainx, 1926). But, for the purpose of stimulating algal cells to rapid multiplication, only a small number of substances come into consideration. Three groups of compounds—carbohydrates, salts of organic acids, peptones—are of primary importance. Even so there are numerous possibilities, but only a few of them are actually involved.

Wherever carbohydrates are beneficial to growth, optimum results are obtained with dextrose, which can be used in concentrations ranging from 0·2 to 5 %. Dextrose should not be autoclaved in acid or alkaline solutions. To avoid decomposition an aqueous solution is sterilized separately from the rest of the medium and, after mixing the two, heat is applied only for a few minutes to kill possible germs from the air.

Acetates are the most important organic salts for the nutrition of algae. For many species of Cryptomonadaceae, Chlamydomonadaceae, Euglenineae, Chlorococcaceae and Ulotrichaceae acetates are as effective or even better than dextrose. It seems that no fatty acid or any other organic acid is the equal of acetic acid. Not even lactic acid, which has so often proved an excellent source of energy for fungi and bacteria, is effective with any alga, so far as I know. Acetates are usually best used in concentrations of 0·1–0·5 %.

Sodium and potassium acetate are alkaline; to prepare a neutral medium acetic acid can be added.

Peptones result from the enzymatic decomposition of proteins. They vary in their nature and action with the variety of proteins and enzymes used. Apart from including mixtures of proteins, polypeptides and amino-acids, they probably contain small quantities of unknown compounds which further the growth of micro-organisms. This fact, and the varying degree of decomposition of the original protein molecule, may be the reasons why various products sold under the name of peptones often produce quite different results. Some peptones are valuable for most purposes, while others, such as the peptone of Witte, so much employed in bacterio-logy, are of little use in cultivating algae. Apart from certain French products (peptone Chapoteau, peptone Vaillant), those manufactured by the American Difco laboratories and, among British products, the bacteriological peptone of Evans Sons, Lescher and Webb are of value.

Peptones are best supplied in concentrations of 0·02–0·2 %, if some other organic substance, such as sugar or acetate, is present, but when used alone higher concentrations are often better. The usual peptones, prepared from muscle, contain sufficient mineral substances, so that further additions are usually unnecessary. Such additions, of course, do no harm and the media used by various workers contain quite a number of chemical compounds, but I have never found them superior to the simpler recipes.

Adjustment to a suitable pH is not quite as essential in media containing peptone and other organic substances as in mineral solutions, but it has to be kept in mind.

B. *Extracts of plant and animal tissues*. Extracts of seeds, especially when germinating, contain many valuable nutritive substances. During germination starch is transformed into dextrine, maltose and dextrose, proteins into albumoses and amino-acids, all of which are soluble in water and can be extracted by boiling. Such seeds appear also to contain subsidiary substances, capable of promoting growth, which are still insufficiently known. Although not suitable for exact experiments on nutrition, such extracts are very valuable for growing cultures of micro-organisms.

Most usually malt is employed, either in the form of beer wort or as malt extract. Concentrations of from 0·2 to 3 % of the extract as sold on the market are suitable. Such extracts are acid and their pH generally requires to be adjusted to one nearer the neutral point. Malt agar is very valuable for growing mass cultures of many Chlorococcaceae and such Ulotrichaceae as *Hormidium* and *Stichococcus*. It is cheap and provides a heavy growth.

When dealing with malt extract and other viscous substances, a piece of paper is cut into halves, one being placed in each scale-pan of the balance. After weighing the extract on the paper, the half bearing the substance is put into hot water and removed as soon as solution of the extract is complete.

Another extract often used in microbiology is beef extract, which contains no sugar, but is of even wider application than malt extract, being mostly used in concentrations from 0·05 to 0·5 %. The best is the standardized product of Difco laboratories which is sold in tin tubes so that it does not dry out, and a suitable concentration once selected can be prepared again and again. The method of weighing and pH adjustment is as for malt extract. When rendered alkaline, in the presence of a surplus of Ca, a precipitate is formed. Beef extract contains peptone, but growth is often strikingly improved by adding more peptone.

Next comes yeast extract, which is almost as generally applicable as beef extract. Difco sells a good preparation in tin tubes. Marmite, yeastrel and other brands, which are cheaper, can also be used. A yeast autolysate giving reliable results can be prepared in the following way: 3 oz. (85 g.) of baker's yeast and 400 c.c. of tap water are well stirred together and placed in a bottle, with the addition of a small quantity of toluol to cover the surface to a depth of a few mm. After shaking in a well-stoppered bottle, this is kept in an incubator at 30–40° C. or in a warm place for several days. When needed, the quantity required is taken up with a pipette, the superficial toluol layer being avoided. Slight traces of toluol dissolved in the watery extract evaporate during sterilization. Yeast autolysate contains vitamins, which may promote growth if autoclaving is avoided or reduced to a minimum. This I would advocate, since a temperature of 120° C. at a pressure of about 2 atmospheres kills even hard bacterial spores in a few minutes. More prolonged heating or heating at a higher temperature can therefore only be detrimental.

Extracts other than those mentioned play no great role in the cultivation of algae and flagellates. Extracts of hay, potatoes, peas and other vegetables may prove worth testing.

5. GELATINOUS MEDIA

Algae living in aqueous or moist habitats can be grown in liquid nutrient solutions or on damp solid substrata. Solid translucent media are often of great value or even indispensable. Gels of diverse nature are used, the most important being gelatine, agar and silica, which were employed in the order given.

Such gel-like media are used for plating, for maintaining cultures and for experimental purposes. The principal object attained by using these media is to fix cells at specified places so that they can be easily observed and can develop into genetically homogeneous populations. Such gels must therefore be sufficiently solid to prevent mingling of the cells, be translucent to admit of observation with the naked eye and with optical instruments, and must contain food substances to allow of multiplication. Owing to the absence of translucency, soil, sand, clay, moist earthenware fragments, etc., are far less suitable than the semi-solid gels. The substances that afford gels are nearly always unsuitable to serve as nutritive substances. An appropriate medium is obtained by mixing aqueous solutions of gels and a suitable nutrient medium.

A. *Gelatine* is no longer employed as much as in the early days of microbiology, but it is still used for various purposes, especially to test organisms for the presence of proteolytic enzymes.

The preparation of a gelatine medium is in no way as troublesome as would appear from the extant recipes, according to which gelatine will not withstand autoclaving and must be sterilized in a steam chamber on several successive days in order to kill bacterial spores; if then it is insufficiently clear, it must be filtered with the help of a hot-water funnel, which consumes time and results in the loss of a considerable part of the medium. Tubes and cotton wool plugs, we are told, must be heated in a dry sterilizing apparatus before use. Even with these precautions the result is not reliable, so that preparatory tests for the absence of bacteria are recommended.

Actually the preparation of a gelatine medium is rather simple, if the stiffening gel, after the addition of the nutritive substances, is not heated to a temperature higher than 100° C. nor for longer than 20 min. Gelatine and nutritive substances can be autoclaved separately in aqueous solution, without any detriment to the elastic and nutritive properties of the final medium.

First, the quantities of distilled water, gelatine and nutritive substances to make up the amount of medium wanted are worked out: 8–10 % of gelatine are needed to produce an elastic gel. The following procedure should then be followed:

(1) The necessary amount of gelatine is soaked in 90 % of the water. (2) The nutritive substances are dissolved in the remaining 10 % of the water and, if necessary, the solution is filtered. (3) To obtain a clear gel, a little dry egg albumen is shaken up with a small quantity of the water and the clear fluid poured into the nutritive solution. The object is to precipitate the turbid particles of the gelatine together with the albumen, when it is heated after mixing with the nutritive solution containing the albumen. A good quality gelatine should, however, give a completely translucent medium without extra clearing. (4) The required number of test-tubes, provided with cotton wool plugs, is prepared. These, as well as the flasks containing (1) and (2), are autoclaved at 15–20 lb. (about 1·5–2 atmospheres) for a minute or two. After the autoclave has cooled, the gelatine and nutritive solutions are mixed, and 6–7 c.c. poured into each of the sterile but still hot test-tubes, contamination by touching the rims of the flasks and tubes being avoided. To ensure that no germs from the air contaminate the cultures, the tubes are heated for 15–20 min. in a preheated steam chamber, cooled down, and finally the media are allowed to solidify in an oblique position.

For this and many other purposes a small steam chamber, which can be kept near at hand, is convenient. I use a model in which the inner perforated basket is 18 cm. in diameter and which is fitted with an automatic level connected with a running tap. A small amount of water only is contained in the chamber so that it can be quickly heated (Fig. 1). The larger type of steam chamber is now only used for special purposes.

Filtering is unnecessary. Even if a little precipitate gets into the tubes, it settles during steaming and cooling before the tubes are sloped and in no way interferes with the results.

B. *Agar media.* Here again, by paying heed to the rule that gel and nutritive substances should not be autoclaved

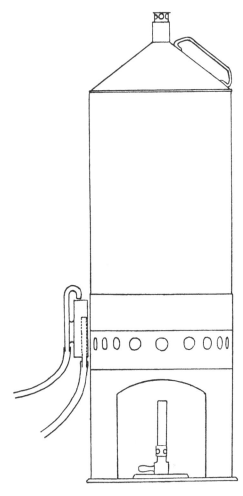

Fig. 1. Steam-chamber with automatic level-regulator. ¼ nat. size.

together, the making of agar media is much simplified and their quality improved. A bad medium is recognized, not by lack of solidification as with gelatine, but by inadequate elasticity. Such a medium easily breaks into pieces when shaken, while a wire loop, such as is used for inoculation, readily penetrates it instead of remaining at the surface. Slopes collapse and the growth of micro-organisms is reduced. Only very ill-prepared agar media remain liquid or purée-like after cooling. It is useless to prepare large stocks of complete agar media in flasks, because the repeated heating necessary for liquefaction is harmful, and if kept in test-tubes for a long time they would dry out. This is not true of a watery agar gel, which can be kept in stock and used for various media.

The procedure is almost the same as for gelatine, the only differences being the amount of agar used and the method of clearing. Much less is needed, slopes requiring 1–1·5 % and plates 2 % of a good quality agar.

Agar media are no longer cleared by filtering, or this is done only exceptionally and after preparatory clearing, when a completely translucent medium is required. In place of albumen, which does not give good results, the agar solution is shaken with a small quantity of cellulose wadding, such as is used in surgery as a substitute for cotton-wool. The requisite amount of agar is heated with distilled water over a Bunsen burner until it dissolves; cellulose is then added and the heating continued for a few minutes. The particles causing turbidity in the agar solution adhere to the cellulose fibres and can readily be removed by allowing the hot solution to percolate through a sheet of the same cellulose inserted in a funnel, or through a woven fabric. The resulting agar gel may not be quite as translucent as after filtration, but is sufficiently clear for most purposes. It should display only a slight bluish opalescence, owing to the Tyndall phenomenon, not a yellowish one, which is a sign that heating has been too prolonged.

The clarified aqueous agar is sterilized in the autoclave, together with a concentrated solution of nutrient substances and the required number of test-tubes plugged with cotton-wool. The autoclaving should not be carried out longer, nor at a higher temperature than is required for killing spores. The pressure is allowed to rise to 2 atmospheres and the source of heat turned off after a minute or two.* After the pressure in the autoclave has gone down, agar and nutrient solutions are mixed and 6–7 c.c. poured into each test-tube without touching or moistening the inner parts of the rims and the plugs. Measuring cylinders or pipettes, if required, are sterilized with the others. For safety the test-tubes filled with the agar medium are heated for 15–20 min. in a steam chamber before sloping.

For plating, the agar media are poured into dry sterilized Petri dishes immediately after removal from the sterilizing apparatus, at a temperature of about 45° C. The usual practice of liquefying stored agar in tubes in a water-bath, prior to plating, is not safe enough for algal cultures, which last longer than bacterial and fungal cultures and might be contaminated by germs from dust settling on the rims and plugs.

The clearing of agar in the way above described can only be carried out if it is adequately diluted. To obtain the

---

* In bacteriological books the heating in the autoclave necessary to sterilize media is given as 15–20 lb. at 120° C. for 20–30 min. When large containers are used, this may well be needed, but for test-tubes and Erlenmeyer flasks up to 500 c.c. capacity such intensive heating is certainly not necessary, because they heat up rapidly. The interior of a large quantity of medium, especially of an agar block in which there are no convection streams, undergoes a much slower rise in temperature than the exterior. As a result the latter is overheated in order to sterilize the inner parts, and the quality deteriorates. For this reason it is advisable to distribute larger quantities into bottles which contain not more than 500 c.c. of medium and are also more handy.

correct concentration, 75 % of the amount of water required is therefore used for dissolving the agar, while a nutrient solution four times the required concentration is prepared in the remaining 25 %. The aqueous stock solutions would thus contain 2 % of agar, when media with 1·5 % agar are required. Most algal nutrient solutions contain low concentrations of nutritive substances, so that stock solutions of four times the required concentration are easily prepared; if necessary they can be filtered.*

It is advisable to wash and soak the dry agar for a day before heating. When the aqueous agar and the filtered nutrient solution are mixed, there is rarely any precipitation. If there is, it usually suffices to allow the vessels to remain for some time in the steam chamber after turning off the source of heat, so that the precipitate can settle. The clear upper portion can be used for plates or for special purposes and the rest for slopes, the tubes being inclined slowly without shaking. A slight cloud remains at the bottom of the tube, but this does not interfere with the results in any way.

When very sensitive organisms are to be cultivated or when experiments on nutrition are made involving the use of agar cultures, the best quality agar is employed and carefully washed for several days, first with tap water, then with distilled water. In order to determine the amount of fluid to be added after washing, the unwashed agar and the required quantity of tap water are placed in a flask and the level marked. After washing, the agar is returned to the same vessel and glass-distilled water is added until the same level is reached. After this the agar is melted and treated in the way described above.

* I have used the same technique for making media for bacteria and fungi, with much saving of time and material. Even 5 % malt extract for fungi, which with the old technique considerably reduces the elasticity of the agar, causes no trouble. This agar is used also for the culture of *Chlorella* and *Prototheca*.

Agar powder, which is very handy, cannot be cleaned in the same way, but must first be converted into an aqueous gel; after this has set it is broken into pieces for washing. This method of cleaning is very efficient. Treatment with acids or putrefaction (Küster, 1913, p. 38; Wettstein, 1921; Kufferath, 1930, pp. 100–3) is liable to reduce the stiffening power of the agar and therefore the elasticity of the medium.

In the rare instances in which glass-like transparency is needed the aqueous agar is filtered in the steam chamber after shaking with cellulose and before the addition of nutritive substances. This is preferable to the use of a hot-water funnel, usually recommended for this purpose. The edge of the filter does not dry and filtering is quicker, because of the more fluid character of the agar owing to the higher temperature. When only small quantities are needed, as will usually be the case, the process is not too troublesome after removing coarse, slimy turbidities.

Employment of agar media more dilute than 1 %, as has been suggested by Wettstein (1921) for growing delicate species, is generally useless. Those that will not grow on ordinary agar media will not do so either on the less stiff ones. Even large species of Desmids and *Spirogyra* divide on 1·5 % agar.

It is sometimes helpful to use, in test-tubes or Erlenmeyer flasks, a very thin, 0·2–0·3 %, mixture which is not rigid enough to form a slope. It has the advantage over liquid media that inoculated cells do not sink to the bottom, where certain algae fail to develop properly. On the other hand, it does not prevent the penetration of filaments or motile cells. This applies to diatoms, Desmids and Cyanophyceae and probably also to other algae.

C. *Silica gel.* The use of silicic-acid gel was first proposed by Marshall Ward (1899, p. 563), but the technique of preparing it was later simplified (Pringsheim, 1914 *a*, p. 57;

1921 *a*, p. 398; 1926 *a*, p. 301) as a result of a suggestion made by Beijerinck (1904, p. 28). A good quality water-glass, though not necessarily a chemically pure one, is treated with hydrochloric acid so that common salt and silicon dioxide are formed. After some time the latter sets as a gel and the solution acquires sufficient rigidity not to flow. For use in cultures the soluble substances present must be removed, since otherwise the salt concentration would be too high. This is usually effected by dialysis through parchment tubes before the solution sets, but the method is troublesome, especially if larger quantities are needed.

Washing can be just as well carried out in the Petri dish in which the gel has set. We dilute pure hydrochloric acid to a specific gravity of 1·1 and add nine times the volume of distilled water. This dilute HCl is mixed with an equal quantity of water-glass (specific gravity 1·08) and well stirred. The mixture is then poured into Petri dishes to a depth of about 3–4 mm., not more than 100 c.c. being prepared at one time and the vessels being thoroughly rinsed immediately after using. When the silica gel has stiffened, the uncovered dishes are placed in a large vessel in which they are washed for not less than 24 hours under a running tap. After that the dishes are inverted so that most of the water drains away. Then they are turned right way up, filled with glass-distilled water and covered with lids. After some hours the distilled water is replaced by a nutritive solution of double concentration, which is left in the dishes overnight and then poured off. Finally the covered dishes are carefully heated in a steam chamber. It is advisable to use a small flame at first and gradually to increase the temperature, since heating at full strength often causes the formation of bubbles within the silica gel. Unless bacterial spores find their way into the dishes after washing, heating for 1 hour is sufficient to sterilize the plates.

Silica-gel plates are used for the culture of autotrophic organisms, which on agar would become overgrown by bacteria. They have proved useful in the preparation of pure cultures of Cyanophyceae (Pringsheim, 1914a, pp. 59, 60; De, 1939, p. 125) and would no doubt be more widely employed, were it realized that their preparation is not especially difficult.

Opaque solid substrata

It is not known who was the first to employ moist solid substrata for growing algae. Earth or bark, moistened and exposed to light, will serve the purpose, but actual cultural work did not commence until the practice of transferring algal material into a new medium, with the object of obtaining subcultures, had been perfected. For such cultures clay, earthenware fragments, etc., were used at an early stage (Klebs, 1896; Küster, 1913, p. 29), but on the whole little further progress has been made. Sterilization can be effected by autoclaving, although it must be remembered that the humus contained in soil and peat may be rendered harmful by heating. Such substrata can therefore scarcely be used for the preparation of bacteria-free cultures. After gentle heating in a steam chamber, soil, peat, clay and bark are useful for preparatory cultures.

For growing mass cultures, or for providing conditions for organisms requiring special treatment, sand, plaster of Paris, filter paper, cellulose, porous earthenware fragments and porcelain constitute valuable substrata (plaster—Pringsheim, 1912, pp. 325 et seq., Pl. IX, fig. 8; 1921a, p. 394: paper—Pringsheim, 1921c, p. 504; 1926a, p. 302). They can be placed in glass vessels, moistened with a liquid medium and sterilized. Their advantage over agar lies in their porous character so that creeping filamentous species can penetrate into the medium. They can also be maintained in a semi-moist condition so that cells inside the substratum are aerated, which is not the case in a liquid medium or in agar. They thus afford a better imitation of natural conditions than can be attained in any other way.

CHAPTER IV

PLATING METHODS FOR MAKING
PURE CULTURES

The aim of any plating method is threefold: (1) to immobilize
single individuals, if possible individual cells, so that they do
not become intermingled; (2) to separate them from other
organisms which might interfere with their development;
(3) to provide them with nutriment and other conditions
favourable for their multiplication into genetically homogeneous populations. Such 'colonies', as these growths are
called, consist of a single species only and, by transference to
a sterile medium, serve as the starting-point for 'pure cultures'. These can be achieved if the individual cells are
isolated from one another sufficiently to admit of their
descendants also remaining isolated and only spreading out
over a small area during multiplication. If an unwanted
organism, present in an inoculum, is able to spread through
or over the medium by its own motile power, the isolation
of the relevant ones for purposes of pure cultures is not to be
expected.

Plating media must therefore be sufficiently solid to prevent mingling of cells and must be capable of absorbing
nutritive substances and water in a dispersed condition.
A third quality, which is often indispensable and at all times
valuable, is translucency of the medium, which admits of
investigation of multiplication and of purity of the growth
with optical instruments and with the naked eye. For this
reason soil, sand or clay are far less suitable than solid colloids
or gels.

1. MIXING

The first cultures of algae were obtained by Beijerinck (1890), using the bacteriological methods devised by Koch in 1881. A nutrient gelatine containing organic substances was liquefied and the algae mixed with it. When it had solidified in a Petri dish, it was exposed to the light for some weeks. The resulting green colonies were picked up with a platinum needle and transferred to fresh media for the preparation of subcultures. Gelatine is no longer used for this purpose, because it is easily liquefied by micro-organisms and because many species fail to grow on it.

Except for the customary usage of agar, the technique for isolating algal cells has not undergone much change. Details are given by Skinner (1932) and Bold (1942, p. 96). After thorough shaking to separate individual cells, the latter dilutes the material with sterile water in several stages, after which a measured quantity is mixed with the liquid agar medium at a temperature of 40–45° in sterile Petri dishes, which are rotated to secure an even distribution of agar and of the algal suspension.

This method of distributing algal cells within the agar has certain drawbacks. It is impossible to know beforehand what will grow and what will not. Some species of *Chlamydomonas* and *Polytoma*, for instance, will not multiply when embedded in agar. Most authors have hitherto been content to isolate the species that will grow best with the method employed, but we are now more ambitious and aim at rearing definite species required for some specific purpose. The colonies that develop inside the agar at the best grow much more slowly than those at the surface, and differ in appearance so that they are not readily recognized.

Conditions inside a solidified agar medium are not favourable to algal growth, which may in part be due to lack of CO_2 which can reach the cells only by diffusion. This, involving thermodynamic molecular movement, is a very slow process and is probably inadequate to supply sufficient CO_2, even when the distance to be traversed is not more than a fraction of a millimetre.

This is one reason why Chodat's method of mixing a suspension of algal cells with an agar medium liquefied in an Erlenmeyer flask can no longer be recommended. In such a solid block the distance from the surface to a particular cell may be considerable. Actually it takes several weeks or months, even under the most favourable conditions, before the colonies develop to such a size that they can be recognized and transferred. Another reason against Chodat's method lies in the impossibility of removing the agar block without breaking it up into small portions, a procedure which is very liable to result in contamination of algal colonies with bacteria present in the medium, but separate from them. It is easy to imagine how often the procedure of raising colonies inside the agar medium must be repeated until pure cultures are achieved, and one may admire Chodat's patience in working with this technique. His difficulties must have been· considerable, as also those of his co-workers and of G. M. Smith (1916), Bristol-Roach (1926, 1927) and others who worked in essentially the same way.

2. SURFACE INOCULATION

For these various reasons it is preferable to spread the algal cells over the surface of the agar after it has solidified (Vischer, 1937, p. 191). Bold (1942, p. 99) holds that this technique 'is not as satisfactory as the plating method,* because bacteria tend to spread abundantly over the surface of the agar'. There is some point in this objection. In certain instances surface inoculation of agar plates is not feasible, because of the development of too many bacterial colonies. But if organisms capable of spreading are present, they will probably extend to the surface anyway, and isolated pure colonies are harder to transfer when embedded than when on

* I.e. that described in the foregoing section.

the surface. In many instances surface inoculation proved far superior when both methods were followed in parallel tests. This is very striking when preparatory purification is undertaken, as is now almost the rule (cf. p. 18). The deciding factor is, however, that delicate forms will grow only with surface inoculation, so that the other method is restricted to more resistant species, which are commonly of far less interest.

For the purpose of such surface inoculation, agar plates, of sufficient consistency (cf. p. 51) and protected as far as possible from contamination, are needed. It is often difficult to prevent infection of Petri dishes by germs from the air, especially by moulds. For long-term cultures, such as are necessary for algae, it is important to avoid the trouble caused by mycelia spreading over the surface. The rapidly formed spores of Aspergillaceae jeopardize the success of a pure culture.

Long-term cultures of micro-organisms in Petri dishes are most suitable for the investigation of their physiological properties, a knowledge of which is often valuable in connection with the technique of isolating pure cultures. The methods are described in every book (e.g. Beijerinck, 1891, p. 781; Küster, 1913, p. 81) dealing with the cultivation of micro-organisms. They comprise tests for enzymes, acidity, etc.

The trouble due to air-borne germs, and to a lesser extent to desiccation, is responsible for the frequent replacement of Petri dishes by other kinds of vessels, mostly flat-bottomed flasks. These can, however, be used only for a few of the purposes for which Petri dishes are suitable, because they do not provide a wide readily accessible surface. Since Petri dishes are indispensable for algal cultures, the following precautions should be taken to avoid infection: (1) The lids should fit closely over the dishes. (2) Both should be sterilized in an oven at about 160° C. for 2 hours, together with a metal sheet with upturned edges to serve as a tray. (3) Before placing the tray with the dishes on the table, clean the latter with a wet cloth. (4) Pour the agar into the dishes immediately after its removal from the sterilizer. If agar from test-tubes is to be employed, it is insufficient to liquefy it in a hot-water bath as often described. Mere heating of the rim of the tube before the agar is poured into the dishes is not always a satisfactory safeguard. The tubes should be heated for at least 30 min. in a steam chamber to liquefy the agar medium and to kill the germs at the rims; after this the tubes are cooled down to about 45–50° C. before pouring the agar into the dishes. (5) Return the dishes on the tray to the cold sterile oven, or, if that is not feasible, leave them on a disinfected surface and cover them with a dust-free bell-jar. (6) The dishes should be brought to the laboratory bench just before inoculation and not kept open longer than necessary. (7) After inoculation the lids should be fastened to the dishes with gummed-paper strips so that they are pressed tightly against them. The whole procedure is not as complicated as it sounds, and will save time, labour and material.

Agar, containing one of the simple mineral solutions, is usually recommended for plating, the agar being carefully washed beforehand. Both practices tend to reduce bacterial growth. Some investigators (cf. Bold, 1942, p. 97) add dextrose to accelerate the multiplication of the algae so that colonies suitable for inoculating subcultures are obtained as soon as possible. But such additions also further the growth of mycelia and are often impracticable, while by no means every algal species grows quicker in the presence of sugar.

If it is desired to promote the growth of algae for plating or other purposes, peat or soil extract is preferable to sugar, but it must be remembered that these may not be heated with agar in the autoclave (cf. p. 49).

The attempt to reduce bacterial growth by using a medium in which such organisms only grow slowly is not of much value. On the contrary, bacteria should become visible by developing into colonies, so that algal colonies which are most remote from bacterial ones may be selected. For this reason an agar medium containing a slight concentration of indifferent organic substances is used (1914 a, p. 60); proteids (i.e. substances formed in the first stages of disintegration of proteins) are most suitable, such as the easily digestible foods prepared for medical purposes. Egg or serum albumen can also be used, though they are not quite so convenient. These do no harm to algae when used in concentrations of o·1–o·4 %, but readily betray the presence of bacteria which form proteinases.

Cells can be distributed on the surface of a solid agar medium so that they shall be isolated from one another by one of three methods, each of which has its special advantages and disadvantages. These are spraying, streaking, and a method that makes use of the faculty of certain algae to creep over the substratum. The best method of dispersing algal cells over an agar surface is to employ a sprayer, but it is rather tiresome to manipulate. The sprayer consists of two glass tubes placed at right angles to one another (Fig. 2). Cotton-wool filters are inserted inside the horizontal part of the tube through which compressed air is blown, and the whole is then autoclaved. The algal suspension should be rather dense and must not contain more than a small number of bacteria (Vischer, 1937, p. 191). A test-tube is used as the container from which the suspension is blown onto the agar plate. It is filled almost to the top. As for other methods of surface inoculation, the agar must have some degree of consistency so that, after taking up the water droplets from the sprayer, it soon dries again. A piece of rubber tubing is connected to the horizontal mouthpiece of the sprayer, the vertical part of which dips into the test-tube. The open, sterile Petri dish is placed vertically and, by blowing into the rubber tubing, a spray is formed which is directed towards the agar surface. Only a very thin layer of liquid

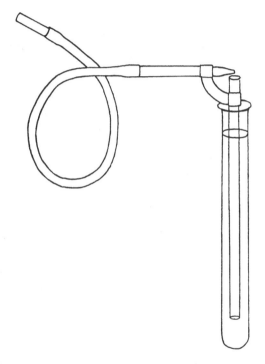

Fig. 2. Spray diffuser. ½ nat. size.

should remain on the surface, otherwise the isolated cells will mingle within it.

The streaking method is specially suitable when few bacteria are present or pre-cleaned material is available. The technique is that used in bacteriological routine. Its aim is to spread the cells over the agar surface so that they are mechanically separated and, under favourable conditions, will multiply to form distinct colonies. For this purpose some force must be applied by means of a suitable tool in order to push the cells over the surface and disperse them as evenly as possible. The bent glass rods used by bacteriologists to distribute germs evenly over the surface are not very suitable for algae unless the latter are of minute dimensions. They are employed to rub a drop of algal mixture over the whole surface of the agar while the Petri dish is slowly rotated. Brushes are handy, but difficult to clean and sterilize. An implement, made by winding a fragment of cotton-wool round one end of a thin wooden rod, is preferable, and can be autoclaved inside a test-tube. Brush or cotton-wool are wetted with the algal suspension and moved in all directions over the agar, contamination being carefully avoided. The effect is similar to that of spraying.

Usually the simplest way is to employ a large loop of platinum or nickel-chromium wire to distribute the cells, which are transferred to the agar either with the help of the loop or by means of a pipette. The surface of the agar must be consistent and elastic and may not be too dry. The wire loop is held at as acute an angle as possible and is gently drawn over the agar surface in a dense zigzag; subsequently the procedure is repeated perpendicularly to the first direction of streaking. An even distribution cannot be achieved with a wire loop, but this is not really disadvantageous. After growth has gone on for some time, plates inoculated in this way show a gradation in density, the colonies being crowded in some and less densely aggregated in other places which are farther from the original point of inoculation; the more widely spaced colonies can be picked up without touching unwanted ones. The final result when a loop is used does not depend to so great an extent on the quantity of cells originally transferred to the plate as it does in the mixing or spraying methods, which afford a more even distribution. As a consequence a single plate, instead of three or four, will suffice to isolate a given species. The method of streaking described in this paragraph is nearly always preferable to the others. In order to ensure obtaining reliable bacteria-free cultures, however, a colony grown on the primary plate should be transferred to a second one and spread out in the same way.

In the preparation of pure subcultures from a homogeneous colony cells are picked up and transferred to a fresh sterile medium. This is done under a binocular microscope. During this process Petri dishes and other vessels must be opened, and are therefore liable to contamination from the air, but the risk is not as great as is usually believed. It can be minimized by using covers which are easily made of aluminium sheet or foil and provided with two apertures for the tubes of the binocular. It is unnecessary to use the sterile inoculation chambers sometimes recommended, which render manipulation much more tedious and, as a result, reduce the number of isolations that can be performed in a given time.

If it is uncertain whether contaminating micro-organisms are air-borne or have been introduced with the original material, some experience can be obtained by studying the flora of plates intentionally exposed to infection by dust. The most frequent organisms introduced from the air are Aspergillaceae, Actinomycetes and certain coccoid bacteria. To know the enemy is halfway to victory, and knowledge of the source of infection almost ensures success in combating it.

The third method of separating algae from bacteria is to make use of the creeping movements of the former, whereby they spread over a sterile agar surface. It is naturally restricted to species displaying such movements. These may, on the other hand, prove very objectionable when the creeping species of a mixed population are not those which are to be isolated (cf. p. 59). Creeping movements are exhibited by Cyanophyceae, Desmids and Diatoms, as well as by the amoeboid stages of Chrysophyceae and other organisms. In many of the members of this biological group creeping is unfortunately associated with the formation of mucilage, which is often accompanied by a multitude of bacteria which are carried over the agar surface.

Creeping movements may, however, sometimes be very useful, either as an additional or even as the only means of obtaining pure cultures. For this purpose the cells or filaments are transferred to the solid surface or dispersed in a layer of melted agar. During multiplication they spread centrifugally over the surface or inside the agar and the bacteria are left behind, so that the individuals at the edge of the more or less circular colony thus formed can be transferred to a second plate to be tested for purity; if this has not been achieved on the first plate the second will afford a closer approximation.

The success of this method depends on the kind of organism and on the rapidity of movement. It is not of much use with Desmids, but helpful with hormogonia of Cyanophyceae and very valuable for small Diatoms. Minute forms, like *Nitzschia palea*, *N. minuscula*, and *N. putrida* and small species of *Navicula*, are easily freed of bacteria in this way (O. Richter, 1903, 1906, 1909; Meinhold, 1911). Cultures of *Nitzschia putrida*, a colourless marine Diatom living on decaying seaweeds, are generally obtained when a small quantity of mucus is put on the surface of agar, made up with natural or artificial sea water. It is, however, advisable to subject the material to preparatory cleaning with a capillary pipette to prevent the cells from becoming overgrown by bacteria or amoebae before they can free themselves by their movements.

The same precaution should be adopted in the preparation of pure cultures of other creeping organisms, such as *Euglena deses*, Desmids or Cyanophyceae. With the last mentioned, much patience is needed to achieve satisfactory results. Cleaning must be repeated several times and, even then, there is no certainty of success, because this appears to depend on the nature of the bacterial flora present in the original material (Pringsheim, 1912; De, 1939; cf. p. 94). In some amoeboid forms, e.g. the colourless flagellate *Naegleria* which is mostly found in the rhizopodial state, this method of purification renders multiplication impossible, because this cannot take place unless bacteria are available as food. Cysts are then formed on the surface of the agar, which germinate only when suitable bacteria are added.

Analogous opportunities to those provided by creeping organisms are furnished by rapidly growing forms. These can sometimes be freed from bacteria by transferring fragments of filaments or groups of cells which have spread away from the original centre of inoculation on to the surface of a suitable agar medium. Moss protonemata, for example, can be purified in this way, if aseptic spores are not available (Pringsheim, 1921 c, pp. 502–3), and this method will probably prove useful also in dealing with branched algae. Ulotrichaceae, Zygnemaceae and filamentous Heterokontae can, however, scarcely be purified in this way, because they glide over the surface of the agar and carry bacteria with them. This is due to their diffuse growth in contrast to the apical growth of protonemata.

When using this technique, special attention must be paid to the method of picking up cells or threads for starting subcultures. The same procedure is useful with every kind of plating, so it is given in detail.

3. TRANSFERENCE OF BACTERIA-FREE CELLS TO STERILE MEDIA

It is not sufficient to separate algal cells from bacteria; they must also be transferred to a fresh medium so that they may give rise to a population called a pure culture. This latter task is by no means always easy. Three instruments can be used for this purpose, viz.:

(1) A wire needle, to the tip of which water films or mucilage containing algal cells will adhere. (2) A small lancet to cut out pieces of agar containing algal cells. (3) A capillary pipette of very fine dimensions for sucking up liquid with algal cells. Transference must always be carried out under a microscope protected from dust (cf. p. 65). It does not take long to acquire the necessary skill, if the correct procedure is followed.

(1) Most workers use a platinum or nickel-chromium wire to pick up algal cells for inoculation into new media. The traditional platinum wire or its substitute is, however, not suitable for most algae. It is only when the colonies are widely separated that there is little or no danger of mixing bacteria-free cells with adjacent bacteria. The lifting of the algal cells can scarcely be carried out without some slight pushing movement, whereby water which may contain bacteria is squeezed out of the agar. There is, moreover, the difficulty of ensuring whether cells have been removed from a small colony, the outline of which is obscured by touching it with so coarse an instrument. Lastly, cells adhering to the wire are not readily detected.

As a consequence I only rarely use wire needles, employing them only in the preparation of subcultures from pure cultures. Even then a loop is preferable to a needle, because for subculturing it is necessary to transfer more material of algae than of bacteria.

(2) A lancet which can be flamed is helpful. It can be prepared from wire made of an alloy of platinum and iridium, being firmer, more elastic and harder than plain platinum and is not so soon roughened by heating as are nickel alloys. A minute blade is fashioned by hammering the tip, and this is sharpened on a stone. It is suitable for cutting out particles of agar, though severance of filaments is not always successful.

(3) The best instrument is generally a capillary pipette, like that used for lifting cells from a liquid medium (cf. pp. 20, 73). This can be employed also on an agar surface. With its help it is often possible to pick up single cells of Diatoms, Euglenas, Cryptomonads, etc., or fragments of filamentous forms. The manipulation can be watched under the microscope, and the successful collection of cells inside the capillary controlled, which is impossible when a wire is used. By exerting pressure on the rubber cap the contents are blown out and thus transferred to a liquid medium. Air bubbles, following the water column, indicate the complete removal of the liquid with the contained cells from the pipette and show that nothing is lost. Such bubbles can be observed also when transference is made to another agar surface, but, when slopes are used, it is best to inoculate into the 'condensation' or extruded water at the bottom of the slope.

It is astonishing how closely colonies can be packed on an agar surface without impeding the selection of minute colonies by this method. The moment the tip of the capillary touches the colony cells are sucked in, often without causing lesion of the agar surface or disturbance to the rest of the colony and to the adjacent area. These facts greatly facilitate control of the process and render success much easier of attainment than with a needle.

After growth has started in a liquid medium or on an agar

slope, its purity is tested for by inoculating some of it into solutions which would favour the development of such bacteria as are suspected of being present. For this purpose a large wire loop is preferable to a capillary, because a greater quantity of inoculum can be transferred in this way, so that growth is quicker.

For such tests the media employed in routine bacteriological work are used, but for reasons of economy in lower concentrations. The medium may, for instance, contain 0·5 % dextrose, 0·2 % beef or yeast extract and 0·2 % of a bacteriological peptone like Difco-bacto-peptone; if necessary, it is neutralized. It can be applied in the liquid form or in an agar medium. The suspicion that bacteria, which will not grow in such a medium, might nevertheless be present, has no practical foundation or experimental proof, but for safety cultures should be investigated microscopically. In such cases it is best to use hanging drops and to examine with an oil immersion. Staining has no special advantages, but can be undertaken in case of doubt to reveal germs that may have been overlooked.

CHAPTER V

THE PIPETTING OR WASHING METHOD

When I started physiological research on Algae thirty-five years ago, the only available method of obtaining pure cultures was by plating algal mixtures suspended in liquid agar. The first bacteria-free cultures of colourless Flagellata, those of *Polytoma*, were prepared in this way (Pringsheim, 1920, 1921). It was fortunate that *Polytoma*, unlike most other apochromatic flagellates, grows well on a suitable agar medium. The earlier pure cultures of green flagellates were all prepared by the plating method, so that those which failed to grow on agar could not be obtained in pure culture (cf. e.g. Mainx, 1927 *b*, p. 348).

A great forward step was made when Lwoff (1929) introduced the method of washing single cells with the help of pipettes, a method previously used by him (1923), as well as by American (cf. Needham *et al.* 1937, Calkins and Summers, 1941) and French scientists, for isolating ciliates. Mainx (1927 *b*, p. 327) already occasionally employed this technique, but did not recommend it owing to its uncertainty. Although fine pipettes had been used for separating unicellular organisms and filaments by Winogradsky and Klebs as early as 1896, many years elapsed before the method was adopted for obtaining bacteria-free material of algae.

The washing method, which in its early years was rather troublesome and often not applicable, has been considerably improved (Pringsheim, 1921 *a*, p. 402; 1936 *b*, p. 51; 1937, p. 639) by the adoption of a number of simple devices. It consists in picking up single cells with micro-pipettes and transferring them to a sterile fluid, whereby most unwanted organisms are eliminated. If the process is repeated several times, the bacterial population is reduced to such an extent that eventually inoculation of a single algal individual into a suitable medium gives rise to a pure culture.

A binocular microscope affording a magnification of 40–80 diameters is used in the detection and manipulation of the algal cells. Contamination of open dishes is prevented by using the metal screen described on p. 65.

Only a small amount of sterile fluid can be used in the process of washing, otherwise it would be difficult to find the algal cells. Dilution of bacteria, on the other hand, is the more effective the larger the volume of fluid employed. Lwoff (1932, p. 14) used single drops on slides, while I work with five to eight drops in watch-glasses. As a consequence four to six washings suffice, as compared with the fifteen found necessary in Lwoff's investigations on *Polytoma* (1932, p. 15), and manipulation is easier with the larger amount of fluid.

The watch-glasses should not be too shallow and should be of the rounded type so that the fluid collects in the centre. Hollow slides afford too small a cavity. The watch-glasses are placed in covered Petri dishes, supported by wire triangles (Fig. 3) in order to prevent slipping or rotation, and are sterilized in an oven for at least $1\frac{1}{2}$ hours at 160–170° C. If stored for a longer period, the watch-glasses, even though enclosed in Petri dishes, must be protected from dust.

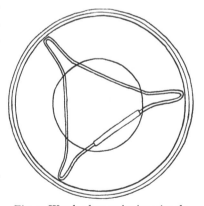

Fig. 3. Watch-glass and wire triangle in Petri dish. $\frac{1}{2}$ nat. size.

Water is often unsatisfactory for washing, since flagellates and zoospores tend to come to rest when transferred from a culture fluid into ordinary water. The use of the culture medium itself for washing is often inconvenient, because solutions containing beef extract, peptone, etc., become foamy through the introduction of air bubbles with the algal cells when these are squirted into the liquid. In most instances balanced mineral solutions or soil extract of a suitable pH will serve the purpose. Even so, it may happen that the algal cells suffer damage during the washing treatment so that their motility gradually diminishes. This serves as a warning that multiplication may fail to take place in the nutritive medium. Unless the capillary is too fine, damage due to mechanical injury scarcely ever occurs.

The process is of course not always accomplished without difficulty. Now and then there are disappointments, resulting from harmful changes as the cells are transferred from one medium to another. Failure may be due to the osmotic concentration, to the deleterious effect of one of the components of a solution even though present in very low concentrations (Provasoli, 1937–8, p. 34), or to an unsuitable pH or gas-content of the solution. It is therefore necessary, when dealing with delicate organisms, to test a variety of solutions till one is found which is not harmful. Failure is, however, rare if due attention is paid to ecological conditions.

The pipettes for transferring fluid into watch-glasses, as well as the micro-pipettes, are made from ordinary glass tubing of about 0·4 cm. bore. Lengths of about 35 cm. are heated in the middle and this portion is drawn out to 6–10 cm., so as to reduce the bore to 0·06–0·1 cm.; when broken across, two pipettes are produced. A little cotton-wool is introduced into the wide end, while just below it sufficient tissue paper is wound around the pipette to form a plug that will fit into a test-tube so that, after autoclaving, the interior of the pipette and a large part of its outer surface are protected from contamination (Fig. 4).

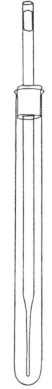

When everything is ready, such a pipette is heated over a micro-burner near its tip and, after removal from the flame, this part is drawn out into a very fine capillary of o·o8 to o·16 mm. bore with the help of a pointed forceps. After again sterilizing the forceps by brief heating over the flame, the part of the pipette thus drawn out is gripped at about 5 cm. from the tip of the latter. A slight but sudden longitudinal pull suffices to detach the end portion, leaving the rest of the original pipette drawn out into a fine capillary tube with a circular opening which is free of contamination. The same original pipette can be drawn out several times so that a single one suffices for the whole series of washings.

The sucking in of the algal cells is effected by capillarity, while their extrusion is accomplished with the help of a piece of pressure tubing placed over the wider end and closed with a glass plug. Ordinary rubber caps are not strong enough to overcome the pronounced capillary forces. In order to take up individual cells or single fragments of filamentous forms the following procedure is adopted. The first watch-glass lodged in its Petri dish and containing a little fluid, including

Fig. 4. Sterile pipette in test-tube. ½ nat. size.

algae, is placed under the microscope, and the lid of the Petri dish is removed. When a suitable cell sufficiently distant from others is observed, the aperture of the capillary is brought near it until it is sucked in. The pipette is immediately removed so that it takes up no other cells and as little fluid as possible. When the aim is to obtain bacteria-free cultures, it is best to pick up more than one cell at a time, the capillary being lowered and raised each time an individual cell has been collected until suction ceases to be effective.

When the capillary is full, the cells are blown into the next watch-glass containing sterile fluid. I do not pass single cells through the various stages of washing, as Lwoff does (1932, p. 14), but deal with ten to twenty or more at the same time. Although a few may be lost, the remainder suffice to inoculate several media after transference to the fifth or sixth watch-glass has taken place.

If the original material is known to be genetically homogeneous, two washed cells are introduced into each culture medium to ensure the occurrence of growth, even if one of the cells is damaged or lost. In nearly every case, however, a suitable medium containing a single cell gives rise to a culture.

A number of special difficulties may be mentioned. Even when cells lie isolated, adhesion to the surface of the glass may interfere with pipetting. This is often so with Cyanophyceae, Diatoms, Desmids, Euglenineae and amoeboid stages. Such cells or filaments should not be dislodged with the tip of the capillary, which would damage them. Adhesion must be overcome by the suction of fluid into the pipette.

Free-swimming cells although undamaged are sometimes apparently lost, which is mostly due to phototaxis. By rotating the watch-glass through 180° the reaction is reversed, and the cells can easily be picked up as they move to the opposite side.

Successful manipulation, unfortunately, does not always lead to the attainment of cultures. It may happen that multiplication fails to take place, even in a great variety of media. This is often experienced with both green and colourless Euglenineae, with a few of the Volvocales and with Cyanophyceae, while with Cryptomonads and Dinophyceae it is the rule. Such experiences warrant the statement that multiplication is often harder to achieve than isolation from bacteria and other unwanted micro-organisms.[*]

As previously mentioned, the possibility of isolation by this technique is not as much determined by the size of the organism as might be supposed. Kufferath (1930, p. 112) was of the opinion that only relatively large organisms—he mentions *Closterium*, *Micrasterias*, *Euastrum*, *Volvox*—could be isolated with a pipette. Actually flagellates less than one-tenth of the length of these are easily picked up, for instance, *Cyathomonas* (12–15 μ), *Chroomonas* (14–16 μ), etc. In the case of non-motile cells the limit may be a little higher, which is chiefly due to the fact that they are not so readily recognized. Organisms larger than 100 μ are not easily collected by capillarity, because wider pipettes must then be used. If necessary, greater suction must be applied, and more extended washings may be necessary to remove bacteria.

It is justifiable to ask why agar should be used at all, if purification from bacteria can be attained by washing. In nature algae do not grow on agar, and its use occasions some trouble, but in actual fact it is often very helpful and sometimes indispensable.

For small organisms, like *Chlorella*, *Scenedesmus* and others, which cannot easily be isolated under the binocular micro-

[*] Schramm (1914, p. 42) already arrived at this conclusion. He says: 'It should be pointed out that it is not the isolation technique which is at fault, but rather the cultural methods. Zoospores and zygospores respectively (in *Vaucheria*, *Oedogonium* and *Spirogyra*), free from other organisms, were obtained—but failed to develop.'

scope, agar provides the best means of separating individuals so that they develop into distinct colonies. Moreover, many species, which exhibit little multiplication in a liquid medium, grow readily on agar, a feature which is not surprising in those which usually grow on the moist surfaces of tree trunks, rocks, mud and sòil. Other species, however, exhibit the like behaviour. There is scarcely an alga which will not grow on agar, apart from certain delicate flagellates (cf. p. 54).

The colonies, which many algae form on agar, are of a characteristic type and display definite morphological features; they may also provide ecological data. In filamentous and especially branching forms, the development of the thallus can be observed more clearly on agar than is possible in a liquid medium. Cell division and creeping movements, sometimes even reproduction, can be directly watched under the microscope. Agar plates, supplied with special additions, such as milk, starch, sugar or containing indicators, are of great value for the study of enzymic processes (Beijerinck, 1891). Agar cultures are also a very convenient means of maintaining strains over a long period of time. Most species remain healthy, and contamination is more readily recognized than in liquid media. For these various reasons agar media continue to be employed for the culture of algae.

CHAPTER VI

BRIEF RÉSUMÉ OF THE MODES OF PREPARATION OF PURE CULTURES

It will be useful to recapitulate briefly the whole process of preparing pure cultures, to facilitate selection from among the various possibilities.

The first point to be taken into consideration is whether the species concerned is one which is motile during the greater part of its life (i.e. is one of the flagellates), or develops flagella for short periods only or not at all, or alternatively whether it exhibits active creeping movements.

In the first case the pipetting method is the most useful, and may be indispensable if the species fails to grow on agar. If it will grow there, pipetting and plating can be combined. With true algae zoospores should be used wherever obtainable and treated like flagellates, i.e. first washed and then plated. Forms possessed of creeping movements are also washed and plated, but it is best not to distribute the cells uniformly, but to deposit them at definite places on the agar.

It is more difficult to purify forms that only show slow creeping movements and those which possess no motile stages whatsoever. If the material contains few other organisms and it is possible to proceed directly to plating, it may be best to transfer single cells or filaments on to agar in dishes or even on to agar slopes in test-tubes. By good fortune, one or other of them may be free of bacteria (Czurda, 1926).

With most algae success is quickly achieved and with the least expenditure of labour by streaking the washed cells over an agar plate. When the required species is contaminated with many other organisms, it is best to inoculate single cells into soil-and-water cultures, the pH of which is adjusted to that prevailing in the original habitat. When sufficient multiplication has taken place, cells are plated on to agar, either directly or, better, after renewed washing. Flagellate forms and zoospores are washed carefully and repeatedly and can then be used to inoculate slopes or liquid media.

Unialgal, preferably soil-and-water, cultures, preparatory to freeing the material of bacteria, have the following advantages: (1) There is considerable certainty of securing growth even of single or widely scattered individuals. (2) These cultures serve for identification and admit of the investigation of morphology and reproduction under the most favourable circumstances, without the possibility of confusion with other algae. (3) Plentiful and healthy material for starting subcultures of all kinds is obtained. (4) Time is available for the preparation of agar plates and other media, and the elimination of bacteria can be undertaken at leisure and when convenient. In this way there is no risk of losing species of interest, as often happens if the original sample is kept in the laboratory for more than a day or two. (5) Plating is effected with a single species and not with a mixture of different algae. The separation of the purification process into two stages, the first involving the preparation of unialgal or species-pure cultures, the second that of bacteria-free or absolutely pure cultures, is very helpful. There is no doubt as to the specific identity of the colonies; hence only those which are likely to be bacteria-free need be selected. (6) Washing can be undertaken at a magnification at which the species could often not be certainly recognized. Even flagellate stages not exceeding 12 μ in length can thus be separated from bacteria. (7) One of the greatest aids in plating lies in the elimination of unwanted organisms, other than bacteria, which might spread over the surface of the agar. Such are fungal mycelia, Amoebae, Diatoms and Cyano-phyceae; which often create so much trouble that the operation is abandoned as hopeless. (8) By preparatory cleaning of the algal material for inoculation the number of plates is reduced and material and time are saved.

CHAPTER VII

TREATMENT AND UTILIZATION OF CULTURES

1. ILLUMINATION

It has long been known that direct sunshine is harmful to algal cultures. Under natural conditions the full rays of the sun rarely fall on an alga and a few centimetres of interposed water are sufficient to reduce the harmful effects. The associated water weeds, turbidity due to floating mineral or organic particles, dissolved humus substances and movements of the surface of the water afford additional protection. Only terrestrial forms inhabiting rocks, tree trunks, etc., are sometimes exposed to full sunlight, but most of them live in shady places and form resting stages when exposed to the sun.

The predominant occurrence of algae in shaded habitats explains why most of them prosper best when grown in diminished light. Hence cultures should be placed near north windows. If these are not available, windows facing east or west should be used, but these should be protected from direct sunshine by panes of ground glass or screens of tissue paper. Owing to the rise of temperature behind such screens, windows temporarily exposed to the sun are not as satisfactory as north windows.

The reasons why sunshine is harmful are not fully clear. Detailed experiments on the actual effects have not yet been undertaken.

The incidence of radiation on the surface of algal cells may have diverse consequences. In a watery medium, as well as on a moist surface, evaporation cannot play the same role as in higher plants. The short rays affect the living substance deleteriously, although ultraviolet rays are excluded by the glass walls of culture vessels. To eliminate the short rays of the visible spectrum a dilute solution of picric acid or potassium chromate is effective, but it is not known whether such light filters are favourable to cultures.

Every object exposed to light rays attains a temperature higher than that of the surrounding water or air, but the medium also gets warmed, the more so as small quantities are usually involved.

It is known that many algae do not survive a rise of temperature and thrive only in cold waters, although the physiological basis for these ecological adaptations has not yet been investigated. The biochemical changes inside living cells involve both synthesis and decomposition, and the balance of the chemical reactions may be disturbed by radiation or rise of temperature.

Two other factors, one biological, one physical, may also come into play. Bacteria, fungi and protozoa multiply more rapidly at higher than at lower temperatures, thus affecting the conditions of life of the algae. In pure cultures algae are actually less affected by rise of temperature than when in competition with other organisms.

The physical factor probably depends on the fact .that gaseous substances are less soluble at high than at low temperatures. The relative pressures of carbon dioxide and oxygen may be reduced to below those necessary for healthy development, a hypothesis which should be tested. Ecological observations show that many algae growing in cold water also prefer fast-flowing rivers or coastal areas, where the water is well aerated. They quickly decay when transferred to small vessels, even if the temperature undergoes no change. Aeration and low temperature act in the same direction. Low temperatures alone are therefore not sufficient to support the cultural growth of species from cold streams. The

medium must be aerated, and the light must pass through a
screen of water. Other cold-water forms, however, are found
even beneath a covering of ice, where aeration is reduced.

The intensity of natural daylight is very changeable and,
during the cold season, insufficient for many species, so that
artificial light must be supplied. Such light contains relatively more of the longer wave-lengths than sunlight. The
heating effect is therefore too strong as compared with the
utilization of light in carbon dioxide assimilation, and cultures
fail to thrive satisfactorily. Moreover, drying out of the
media causes trouble. Such difficulties can be overcome by
filtering the light through a layer of cold water, about 10 cm.
thick. It is cheaper to surround the electric bulb on all sides
with a water screen than to use a rectangular one (Warburg,
1919). Hartmann (1921) fitted two cylindrical vessels con-
centrically, the one inside the other, with tap water running
through the intervening space, while the bulb was suspended
within the inner one. The cultures were grouped around this
apparatus.

Where the tap water is chlorinated such a device remains
in good working order for a long time, but in other localities
green, brown and bluish specks, formed by Chlorophyceae,
Diatoms and Cyanophyceae, soon appear on the glass under
such favourable conditions for growth and prevent sufficient
light from reaching the cultures. In order to remedy this
defect I suggested (1926 a, p. 309) the separation of the water
used for screening and cooling. The space between the glass
cylinders is filled with a dilute solution of copper sulphate
in distilled water. The copper salt not only serves as a poison,
but absorbs infra-red and red rays so that the illumination
is more similar to that of daylight. Such a water screen would
soon be heated to boiling-point, but can be cooled by means
of a coil of copper tubing through which tap water runs. The
water should contain no other metal so as to avoid electric

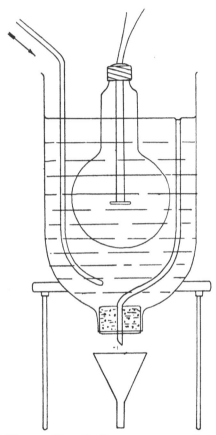

Fig. 5. Direct cooling. Cooling screen surrounding electric bulb. ¼ nat. size.

currents, which would destroy one of the metals. By using the devices illustrated in Figs. 5 and 6, siphons, which may become blocked and cause an overflow, are rendered unnecessary.

When a 500 W. bulb is used, the distance between it and the cultures should not be less than 50 cm. Species that

grow better at lower light intensities are placed behind the
others. Such an illuminating apparatus is very useful for
maintaining strains during the winter. Algae can be grown
in this way in the absence of natural light, at an almost

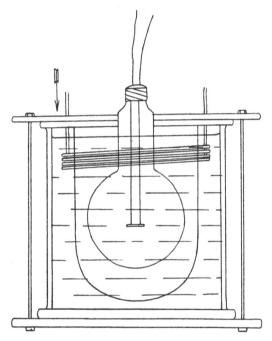

Fig. 6. Indirect cooling. Cooling screen, with coiled copper tubing,
surrounding electric bulb. ¼ nat. size.

constant intensity of illumination. Strains in the culture
collection have thus been maintained for twenty years and
more. Artificial light is, however, at present only a substitute
and never as efficient as daylight. When the brighter season
commences, numerous cultures are always transferred to a
north window where they grow more quickly and present a

healthier appearance. The problem of supplying optimal as well as constant illumination has still to be solved.

In order to expose numerous cultures without mutual shading the test-tubes are suspended in rows in front of the source of light. As first recommended by Richter (1911), the tubes are attached with the help of hooks to wires fixed horizontally across the window in such a way that they do not touch the panes, the temperature of which changes more than that of the surrounding air. A strong, rigid, galvanized wire is recommended, since a thin wire will bend under the weight of the cultures.

Fig. 7. Brass clip for suspending culture-tubes. 1/1 nat. size.   Fig. 8. Wire clip for suspending culture-tubes. 1/1 nat. size.

When using an artificial source of light, curved brass or iron bands are better than wire. But satisfactory devices are provided by wires attached to a wooden framework or by a piece of wide-meshed wire netting bent into the shape of a cylinder (Chu, 1942, p. 286).

A simple means of fastening test-tubes and flasks to their supports is constituted by clips made of elastic brass band, consisting of an open ring and a hook (Fig. 7) (Pringsheim, 1926a, p. 293, fig. 1). Cheaper clips can be made of galvanized wire, 1 mm. thick and about 12 cm. long. For shaping, the wire is wound round a test-tube while the other end is bent into the form of a hook (Fig. 8). Such clips can easily be widened or narrowed to accommodate test-tubes of slightly different widths.

2. THE MAINTENANCE OF CULTURES

When a pure culture has been obtained it should be kept alive, and this not only to avoid having to repeat the process of isolation. Original strains, which constitute type-forms of species or have been used in experiments, should be preserved, as has rightly been emphasized by Vischer (1937, p. 191). He remarks: 'Autoren, die ihr Untersuchungsmaterial der Vernichtung preisgeben, verzichten darauf, dass ihre Publikation jemals nachgeprüft oder ergänzt werden kann und wissenschaftlich ernst genommen wird.'

A few delicate species, however, tend to die out suddenly, e.g. some *Euglenas*, colonial Volvocaceae such as *Eudorina*, Dinophyceae and Chrysophyceae. These are more safely preserved in soil-and-water cultures, from which they can again be isolated in a bacteria-free state, if necessity arises. Most algae, however, give little trouble if certain precautions are taken. The maintenance of most pure cultures is far easier than preparing them, but to avoid the loss of valuable strains, I give the reader the benefit of my experience.

The strains in the culture collection are kept on agar so far as possible, since subcultures are more easily prepared from them than from cultures in liquid media. Moreover, contamination is readily recognized when cultures are examined with a lens. Small tubes, about 12 by 120 mm., are used for routine cultures to save space and material. The quantity of agar should not be too small, but, to avoid rapid desiccation, the surface area of the slope should not be too large.

Subcultures should be inoculated with a fair quantity of algal material, which is allowed to multiply under good conditions of illumination. Before maximum growth is attained the tubes are transferred to a situation with duller light. It is convenient to store a number together in large glass jars, which are covered with paper lids. This protects the cotton-wool plugs from dust and reduces evaporation. A more tightly fitting lid would cause the plug to become damp and encourage the development of moulds, the gravest danger to which cultures are exposed.

Experience shows which species keep for a long time and which require frequent subculturing. Their term of life depends partly on the medium. It is therefore advisable to test a strain in several different media, until the most suitable one has been found.

On the whole Chlorococcaceae, as well as most Volvocales and Ulotrichaceae, keep very well, and subculturing need not be undertaken oftener than four times a year. Others, which are not so long-lived, should be kept apart in special containers. To reduce the danger of losing strains each is maintained in at least three cultures of different ages, the oldest being used for making subcultures, while the others are available in case of misadventure. Tubes of the same strain are held together with rubber bands.

If a culture fails to give rise to subcultures because it has become too dry or because it contains insufficient living material for purposes of inoculation, it can nevertheless usually still be saved by pouring in, under aseptic conditions, a small quantity of a suitable liquid medium sufficient to cover the bottom of the slope. Even if only a few living cells are present, they will commence to multiply after some time. Species of *Botrydium*, *Euglena*, *Polytoma* and others grow better on agar when transferred from a liquid medium and vice versa, so that alternate cultivation in liquid and agar media is advisable.

Instead of keeping sets of cultures in glass jars as recommended above, the individual tubes can be protected. This is the rule with soil-and-water and with putrefaction cultures, which cannot be kept in small tubes. They can be protected in various ways. The rubber caps used by bacteriologists are expensive and, when old, adhere to the glass, especially when exposed to light. Caps of tin or aluminium foil, which are attached by means of insulating tape, are quite as good. Still more appropriate is 'waxed paper' (i.e. paper impregnated with solid paraffin), kept in position by a rubber band or a strip of gummed paper. If a certain degree of permeability is desirable, cellophane or ordinary paper can be employed.

If a special illumination plant (cf. p. 83) is not available during the winter season, cultures should either be provided with additional illumination from ordinary electric bulbs or be kept at a south window or in a greenhouse, protected by tissue paper or some similar screen. Although cultures show little growth under such conditions, they keep rather well if the room is unheated. High temperature combined with low light intensity is dangerous.

When cultures have to be sent through the post, it is safest to seal the tubes. Such cultures should be neither too young nor too old, preferably at the stage of most active multiplication. If possible, they should be dispatched when the temperature is moderate. Most species remain alive under such circumstances, even if they are darkened for days or weeks, but subcultures should be made as soon as possible.

Although for research purposes any culture from which subcultures can be prepared is suitable, those needed for teaching must show characteristic morphological and reproductive features which are not always easily attained. Strains that have been kept on agar for a long time are unsuitable. These take some time to improve, which is best accomplished by transferring them to soil-and-water cultures which more closely fesemble natural conditions than does an agar culture. These cultures, which cannot be sent far, are placed in ordinary tubes which are superficially sterilized.

3. THE USE OF CULTURES

The diverse uses to which algal cultures (especially pure cultures) may be put have repeatedly been discussed (Küster, 1913, pp. 1 et seq.). It will be sufficient to enumerate them with some comments. Cultures of algae are of value for the following purposes:

(1) morphological and cytological studies;
(2) studies of ontogenetic development;
(3) definition of species;
(4) variability and mode of origin of species;
(5) cellular pathology;
(6) cellular physiology;
(7) physiology of development;
(8) physiology of nutrition;
(9) physiology of metabolism;*
(10) physiology of irritability, especially tactic movements;
(11) ecological studies;
(12) biological examination of water;
(13) as a food for animals under investigation.

(3) Algal floras often give the impression that most of the species mentioned are well defined and readily recognized. In many families, however, this is not so, because variation due to environmental influences, or modification as it is better called, cannot, by mere microscopical investigation, always be distinguished from variability due to hereditary differences. This difficulty is greater than is usually admitted, and algal cultures are the best means of overcoming it.

---

* Nutrition and metabolism are mentioned separately, because the investigation of enzyme production and activity and the fate of special compounds within a medium is quite a different matter from the testing of various nutritive solutions in order to attain the best growth.

The capacity to undergo modification varies greatly in different systematic groups, so that not all species described by taxonomists can be regarded as well established. Cyanophyceae, Euglenineae and Chrysomonadineae are certain of the groups in which much confusion has been created by the description of new species, without a sufficient knowledge of those already established and without taking into account their capacity for modification.

(1) and (6) A better knowledge of cellular morphology and physiology could be attained, if healthy cultures of greater homogeneity were used.

(8) and (9) Few species have been investigated in this connection and wider studies are requisite to avoid premature generalizations (Gunderson and Skinner, 1932).

(10) Tactic movements are less well known than tropic ones, which are readily investigated. The material for research on tactic movements has usually been found by mere chance and has remained available only for a short period, so that experiments were often discontinued before definite results were obtained. Repetition, which alone affords reliable conclusions, was impossible. Moreover, cultural material is much more suitable for exact experiments, because it can be raised under homogeneous conditions (Pringsheim and Mainx, 1926; Mainx, 1929 a).

(11) Ecological studies can be promoted in different ways by the culture of algae. By subjecting a certain species to different media, temperatures, illumination, etc., its needs can be defined and compared with those prevailing in the natural habitats; by studying the distribution of a species, in conjunction with ecological data gleaned from its habitat, experience is gained which can be utilized and checked in its culture. As a rule algae growing in water are more suitable for precise ecological investigations than any other group of organisms. The conditions of existence of submerged plants

are more readily controlled than those of terrestrial plants which live partly in a moist porous medium and partly in the air.

(13) Greater use should be made of algal cultures for feeding small animals, since in this way various problems, as yet scarcely broached, may be solved. Such problems concern the essential chemical elements, well investigated in plants, but not in animals; the mode of digestion of certain organic compounds like algal food reserves; the method of trapping and of devouring food particles.

CHAPTER VIII

DATA ON THE CULTURE OF THE VARIOUS TAXONOMIC GROUPS

There is at present little evidence of a relation between the taxonomic position of a species and the conditions under which it lives. Rhodophyceae and Phaeophyceae, it is true, consist in the main of marine forms, but even here there are some exceptions. The few relations that are recognizable are not very obvious or far-reaching.

Diverse algal floras contain data on the ecological relations of certain taxonomic groups (cf. Lemmermann, 1910, and the volumes of Pascher's *Süsswasserflora*). Fritsch (1935) also gives in each section well-balanced ecological considerations, based on observations in the natural habitats.

The indications given in floras as to the conditions prevailing in the habitat, such as katharobic or eutrophic, are valuable, but the description of taxonomic groups or individual species as holophytic or saprophytic* is of little use unless based on experimental proof. What is actually known is seldom more than that the species concerned occurs in more or less pure or in polluted water. The causal link need not, however, be that of nutrition, but may quite well be a capacity to resist influences which are harmful to many other organisms (Pringsheim, 1914 a, p. 86).

Within a taxonomic group adaptation to different ecological

---

* Such statements occur for instance in Pascher's *Süsswasserflora* (Lemmermann, 1913): 'Ernährung holophytisch und saprophytisch' mentioned under Euglenaceae (p. 123), and in the genetic diagnoses of *Lepocinclis* (p. 134), *Phacus* (p. 135) and *Trachelomonas* (p. 142), but not in that of *Euglena*. No references are given, and there is no information as to how the author arrived at these conclusions.

surroundings is more frequent than uniformity of habitat. This is one of the most important considerations to be taken into account when multiplication of a given species is desired. In every order certain species seem to be confined to acid soils or waters, while the majority prefer a neutral or even alkaline reaction, or are indifferent to slight variations of the H˙-ion concentration. This is so among Cyanophyceae, Chrysophyceae, Cryptophyceae, Volvocales, Euglenineae, Chlorophyceae, etc. A large number of Desmidiaceae are acid-loving, while among Chlorococcales, Heterokontae and Diatoms most species seem to prefer an alkaline medium.

Specific adaptation to a certain pH range need not be due to concentrations outside that range interfering with protoplasmic activity, because differences in pH also affect the solubility and therefore the availability of indispensable chemical elements, such as Ca, Mg, P and Fe. Adaptation to temperature has scarcely been investigated in detail, although many isolated observations have been collected. The influence of osmotic concentration is much better known, especially for species found in salt water.

The following pages contain hints on the general conditions of culture in the larger taxonomic groups.

CYANOPHYCEAE

These are very diverse in their ecological adaptation. So far only few species have been grown in culture. As a rule Cyanophyceae are not very sensitive to competition with bacteria and can be grown on agar. It is easy to isolate single cells, hormogonia, spores or fragments of filaments with the capillary pipette and thus to remove other organisms, although the preparation of bacteria-free cultures requires much patience. Many disappointments await the investigator who tries to obtain pure cultures by plating. Areas of a plate, which are seemingly clean, usually prove to contain bacteria when material is transferred from them to sub-cultures on media with organic substances. The frequent creeping movements are of no great help, because the paths traversed by the threads are infected with bacteria. The use of paper, plaster of Paris, bits of flower-pots, etc., on which many blue-green algae flourish, is also of little value in making pure cultures. Cultures of this kind made up with mineral solutions, however, provide living material suitable as a starting-point for further, cleaning.

Silica gels (Pringsheim, 1914 *a*, pp. 57 et seq.; Schramm, 1914, p. 39; De, 1939) have been found helpful, since a small proportion of single filaments transferred from such cultures to agar media often prove to be free from bacteria. But it is not clear whether this practice is really essential and whether it would be effective in all cases. It may well be that pure cultures could be attained in some other way, with the same expenditure of labour. However that may be, the use of silica gel can be recommended. It is also well to transfer from liquid to gelatinous media and vice versa in order to reduce the growth of bacteria.

Blue-green algae have attracted a number of research workers, because of their obvious biological importance. Various devices have been tried to secure pure cultures. Fogg (1942) obtained bacteria-free cultures of *Anabaena cylindrica* by treating the alga with chlorine water (25 p.p.m.) for 2 min. After washing with sterile water, it was plated and found devoid of contaminating organisms. The author was, however, unable to obtain pure cultures of other species by this method and suggests that success in the one instance may have been due to the absence of spore-forming bacteria.

Another promising method is to destroy the bacteria with the help of ultraviolet rays from a mercury vapour lamp. This was first successfully achieved with *Anabaena variabilis* by Allison and Morris (1929; cf. also Allison and Hoover,

1935; and Allison, Hoover and Morris, 1937). Bortels (1940), who adopted the same method, describes it in detail. A small quantity of the alga is dispersed in 2 c.c. of sterile water and exposed for from 1 to 8 min. in a quartz tube to a quartz mercury vapour lamp, placed at a distance of 20 cm. Shaking appears to be essential, otherwise the algae would shade the bacteria and protect them from the ultraviolet rays. So far as practicable, this method is at present the only one capable of eliminating bacteria which adhere to the surface of an alga, unless the latter forms zoospores. It would therefore certainly be worth while to test its value carefully, possibly in combination with oxidizing agents, like Cl, H_2O_2, etc., which seem to have a greater effect on bacteria than on algal cells.

A high concentration of salts hinders the multiplication of bacteria living in water more than that of many Cyanophyceae. I have used as much as 0·5 % KNO_3, although the best conditions for multiplication of the algae must first be determined.

The preparation of pure cultures should be attempted during the favourable season of the year, i.e. from May to September, and a cool, bright room should be chosen for the purpose.

CHRYSOPHYCEAE

Little is known about the cultural conditions favouring the growth of members of this class, which contains many forms thriving at low temperatures. *Ochromonas* was grown by Meyer (1897) in inorganic and dilute organic media, together with the bacteria on which it feeds. Repetition of these experiments has confirmed that elimination of bacteria is followed by failure of the organism to multiply. *Chromulina* is easily grown in media with soil extract, although pure cultures have not yet been obtained. The first member of Chrysophyceae reared in pure culture was *Synura uvella* (Mainx, 1929 *b*); this was accomplished with the help of the pipetting method. I can confirm this result and also that it is comparatively easy to secure multiplication of this species in dilute solutions made of soil extract or of beef extract and acetate. Other members of the class have been cultivated in soil-water cultures, but not in liquid bacteria-free media.

CRYPTOPHYCEAE

Although some species of *Cryptomonas* grow readily on agar (Wettstein, 1921; Reichardt, 1927), pure cultures obtained by plating have not yet been achieved. Pipetting is easy, and motile cells are readily freed from bacteria, but multiplication of such cells is usually unreliable and sporadic, except in the colourless *Chilomonas* (Pringsheim, 1921 *b*, 1935; Glaser and Coria, 1930; Loefer, 1934). In soil-water cultures Cryptomonadineae multiply and thrive, but with many forms it is difficult to obtain active swarming.

DINOPHYCEAE

The first successful attempts to grow Dinophyceae go back to 1908 (Küster), although no pure cultures have been obtained of the colourless marine form concerned. Nothing further was accomplished during the next twenty years, but more recent investigations (Höll, 1928; Lindemann, 1929; Bachrach and Lefèvre, 1930–; Barker, 1935; Köhler-Wieder, 1937; Diwald, 1938) show that multiplication of these delicate organisms can be obtained. Such cultures are useful for taxonomic purposes. Both fresh-water and marine forms have been cultivated, but in view of the immense number of biologically diverse types only the very earliest steps of cultural investigation can be said to have been achieved.

DIATOMS

These organisms, represented by many diverse species, are found in ditches and shallow waters, in streams and the littoral region of lakes, often almost to the exclusion of other algae. It appears therefore that they have certain unknown living conditions in common. Other species, in part belonging to the same genera, are fresh-water planktonts or are found only in the sea, while others again are terrestrial. These are not readily observed, but multiply in soil suspensions made up with water or mineral nutritive solutions.

Many species grow well in soil-water cultures, prepared with tap or sea water, as the case may be. Isolation of single cells is usually possible, but is easier with planktonic than with creeping forms, since the latter tend to adhere to the glass. Soil extract is a valuable ingredient of the cultural medium. On agar the movements of pennate diatoms serve to rid them of bacteria, but in spite of this the culture of diatoms is often troublesome. Certain species will only grow in very dilute solutions in which the number of cells obtained is small. Others grow readily on agar, but not in liquid media. Even a very dilute (0·2–0·3 %) agar gel considerably improves the growth of diatoms. This may be due to the fact that they are prevented from sinking to the bottom, where conditions are unfavourable, although Bachrach (1927) states that even traces of agar, which are insufficient for this purpose, already have a marked influence.

Numerous cultures have been prepared since Miquel (1890-2) produced his first papers on the subject. Richter (1903) claims to have prepared the first bacteria-free cultures, although it appears that Miquel already obtained 'Cultures des Diatomées à l'état de pureté absolue' (cf. Allen and Nelson, 1910, p. 424). Allen and Nelson (1910), Meinhold (1911), Allen (1914) and Wiedling (1941) contributed further to our knowledge of the physiology of Diatoms. Little has since been published on bacteria-free cultures, and our insight into the ecology and physiology of this vast group is still very restricted (cf. Geitler, 1932; Harvey, 1933; Bold, 1942, p. 117 for references).

EUGLENINEAE

Most members of this group are rather erratic and do not multiply rapidly even under the most favourable conditions. *Euglena gracilis* is the species which has been most often studied in cultures and used for physiological investigations (Zumstein, 1900; Ternetz, 1912; Pringsheim, 1913; Mainx, 1927 *b*; Jahn, 1931; Dusi, 1930 *a*, 1931, 1933; Hall, 1931, 1939; Lwoff, 1932), sometimes under the synonym *E. agilis* auct., non Carter (Baker, 1926; Hall and Schönborn, 1939). *E. gracilis* multiplies better than others in artificial cultures, but far less rapidly than certain Volvocales. The same difference exists between the colourless members of the two groups.

On the other hand, Euglenineae keep better than Volvocales and survive conditions unfavourable to multiplication for long periods without of necessity forming special resting stages. The individuals may remain motile for months or even years within test-tubes, provided that desiccation is prevented. Such individuals do not even, so far as can be seen, present many morphological differences from those in active multiplication. This observation has been made on many green species, like *Euglena spirogyra* or *Lepocinclis Steinii*, and on colourless ones like *Astasia linealis* and species of *Menoidium* and *Rhabdomonas*.

Zygotes and true cysts are lacking in the Euglenineae, but palmelloid stages are formed, and these can pass by progressive stages into resting cells with mucilaginous envelopes. When submerged in nutrient solutions these give rise to flagellate cells, which can be washed in preparation for pure cultures if the species multiplies in the medium employed. The apparent contradiction between perseverance and fastidiousness is striking, though not restricted to Euglenineae. The explanation may lie in the food economy, paramylon vanishing slowly under conditions which, in one of the Volvocales, would cause starch to disappear quickly.

It is impossible to give general directions as to the preparation of pure cultures of Euglenineae, although all of them thrive in soil-water cultures. The statements made in many books are misleading and erroneous, because they are generalizations based on the experience gained with *Euglena gracilis*. Zumstein (1900) and Ternetz (1912), in their fundamental papers, give recipes for media containing 1 % peptone Witte and 1 % citric acid, or even higher concentrations of these substances. No one else has succeeded in growing strains of *Euglena gracilis* in these media (Pringsheim, 1913, p. 2; Mainx, 1927 b, p. 321) and there is evidently some mistake. Since the publication of my paper I have tested eight different strains and have found that generally these media admit of multiplication only when diluted at least ten times; even then the results are poor. Better results are obtained with yeast or beef extract (Pringsheim, 1913, p. 30; Mainx, 1927 b, p. 326) and acetic acid, which is best included (Lwoff, 1932, p. 91). Addition of peptone still further improves the medium, and excellent multiplication takes place in a medium containing 0·2 % sodium acetate, 0·2 % beef extract, 0·2% Difco Tryptone or even higher concentrations. The pH may be as low as 5.

Now that really good media are available, they can be applied also to other Euglenineae. The results vary, but most other species do not withstand so high a concentration of acetate and peptone, although beef extract is a good nutrient for many of them, e.g. *E. viridis*, *E. deses*, *E. pisciformis*, *E. anabaena* (Mainx, 1927 b, p. 327). With the exception of

E. mutabilis (syn. *E. Klebsii* Mainx), other species will not thrive in so acid a medium as *E. gracilis* (Dusi, 1930 b). Many species, such as *E. caudata*, *E. sanguinea*, *E. Ehrenbergii*, *E. proxima*, etc., fail to grow in any of the media, although they do very well in soil-water cultures. That is also true of *Phacus*, *Lepocinclis* and *Trachelomonas*.

Among colourless Euglenineae most Astasiaceae are saprophytic, while the Peranemaceae are holozoic. The former multiply abundantly and without exception in starch-soil cultures (Pringsheim, 1937, 1942). Only a few species have hitherto been grown in pure culture, and of these only two (*Astasia longa* and *A. ocellata*) do so readily and without difficulty.

Cultivation on agar has so far proved possible only with the green species raised in pure culture and with *Astasia longa*.

VOLVOCALES

All the members of this group, with the possible exception of certain Polyblepharidaceae and of *Volvox*, seem to grow on agar. Pure cultures can be obtained by concentrating individuals phototactically and distributing some of the material on to the surface of an agar medium containing a dilute mineral nutritive solution. The pipetting method also affords good results. Single cells or coenobia usually multiply quickly in mineral nutritive solutions with soil extract, if the reaction is approximately neutral.

Growth in liquid or on agar media is improved by adding acetate and beef extract, although lower concentrations are suitable for Volvocales than for Euglenineae. Subcultures in liquid media often fail to multiply or persist only in the palmelloid state. Motility can be restored by transference to soil-water cultures.

Species of *Dunaliella*, *Chlamydomonas*, *Carteria*, *Chlorogonium*, *Pteromonas*, *Lobomonas*, *Gonium*, *Pandorina*, *Eudorina*, as well as the colourless genera *Polytomella*, *Polytoma* and *Hyalogonium*, have been grown in pure culture. They are able to live under fairly diverse conditions, but the limits are narrower for the maintenance of swarming than for mere multiplication, while the cells of the species in question will survive under a wide range of circumstances.

A few Volvocales, such as various species of *Chlorogonium*, *Gonium* and *Eudorina*, though showing plentiful multiplication, are rather delicate or short-lived in pure cultures. Sometimes they lose their colour suddenly and disappear. Strains can be preserved in soil-water cultures, from which new bacteria-free subcultures can be prepared as required.

Volvox has been grown in pure culture by Uspenski (1925; Uspenski and Uspenskaja, 1925) in a medium containing a mixture of mineral salts including iron, with citrate to prevent precipitation. By repeated subculturing bacteria were eliminated. According to him *Volvox* does not utilize dextrose, the only organic substance investigated. Other coenobial Volvocaceae, such as *Gonium*, *Pandorina* and *Eudorina*, appear to grow better with peptone or beef extract than with nitrate, but only when these substances are supplied in very low concentrations. Perhaps *Volvox* behaves in the same way.

CHLOROCOCCALES

Species of Chlorococcaceae (e.g. *Chlorococcum*, *Dictyococcus*, *Hypnomonas*, etc.), Chlorellaceae (e.g. *Chlorella*, *Muriella*, etc.), Oocystaceae (e.g. *Chodatella*, *Oocystis*), Selenastraceae (e.g. *Ankistrodesmus*, *Dactylococcus*, *Kirchneriella*), Dictyosphaeriaceae (*Dictyosphaerium*), Hydrodictyaceae (*Pediastrum*, *Sorastrum*) and Coelastraceae (*Coelastrum*, *Crucigenia*, *Fernandinella*, *Scenedesmus*, etc.) all grow readily on agar with mineral solutions. In nature one sometimes finds populations composed almost entirely of a single species, but usually several species live together. If spread over the surface or enclosed in melted agar, mixed populations give rise to colonies, the majority of which belong to one or two of the species predominant in the original material. The less widely represented species mostly get lost in this way.

It is therefore advisable to isolate the rarer species by picking up single cells or colonies with a capillary pipette. The unialgal cultures thus obtained provide material from which plates can be inoculated by streaking. The cells of many Chlorococcales are so small that complete elimination of bacteria is hard to achieve by mere pipetting, while plating without preparatory cleaning is often ineffectual, and pure cultures are most quickly obtained by a combination of both processes. Even so, there is often an element of chance, since bacteria adhere to the cell envelopes which in certain species are somewhat mucilaginous. If swarmers are available these are used for plating, because the colonies to which they give rise are to a larger extent bacteria-free than those derived from non-motile stages. Zoospores are usually obtained when an agar culture, which may contain bacteria, is covered with water or an appropriate liquid medium. In the case of families which do not multiply by motile spores, plating must often be repeated several times. For the first plating an agar medium containing mineral salts only is used, while for the later ones it is better to employ a medium with a low concentration (0·1–0·25 %) of sugar, together with peptone or yeast extract, on which both algae and bacteria grow better.

Pure strains of Chlorococcales are usually easy to maintain. Subcultures can be prepared even from half dry slopes.

To this group belong various symbiotic algae living in animals and lichens. The so-called *Zoochlorellae* are actually species of *Chlorella*.* Limberger (1918) claims to have

* Various other algae occur within the bodies of small aquatic animals, but none of these has so far been cultivated.

isolated the symbionts from *Euspongilla lacustris* and *Castrada viridis*, but his method does not appear to have provided adequate safeguards against adulteration with other algal cells. The *Chlorella* from *Paramaecium bursaria* has been grown in pure culture by Loefer (1936), and I am obliged to him for sending me a culture. The strain is morphologically indistinguishable from other species of *Chlorella*, although physiologically different. The 'gonidia' of lichens belong to the genera *Trebouxia* and *Coccomyxa*. The algal symbiont of *Cladonia* is said to be a *Chlorella*, but that is very doubtful. Chodat (1913, p. 186), Warén (1920) and Jaag (1929, 1931, 1933 *a*, *b*) have isolated many of these forms in pure culture and established their specific differences. As in many other members of the order growth is favoured by sugar and organic nitrogen compounds. For the preparation of pure cultures pieces of the lichen, carefully selected and cleaned, are either crushed or better cut into fine slices with a razor. The algae grow well. Purification by plating is carried out in the same way as in other small non-motile algae.

ULOTRICHALES AND OEDOGONIALES

Certain genera (*Hormidium, Stichococcus*) are commonly grown in pure culture and used for experimental work. Both they and others are easily isolated, when zoospores are available. No special methods are required. Most of the common species of *Oedogonium* appear to be easily isolated in pure culture. Fragments of young plants, not yet overgrown with epiphytes, are cut out under a binocular microscope and transferred to an inorganic nutrient solution with soil extract. When some growth has occurred, material is transferred to a fresh medium and the culture placed in an incubator at 25° C. The new solution, the higher temperature and the darkness all operate together to induce zoospore formation, which sets in after one or two days. After the zoospores have been phototactically assembled, they are transferred to another vessel containing the same medium which must be sterile and at the same temperature. After once again assembling zoospores phototactically, they are used for plating (Mainx, 1931, p. 487). Mainx's method gave inconsistent results, and it might be possible to improve upon it by isolating single zoospores in a sterile medium, instead of mixing them with agar. Subculturing on agar slopes is only possible with species of narrow width. It is best to alternate between liquid and agar media, if it is a question of preserving pure cultures. In liquid media species of *Oedogonium* retain their vitality extremely well and will afford subcultures, even when the material is almost bleached. In a new medium there is immediate formation of zoospores which give rise to many young plants.

SIPHONALES AND SIPHONOCLADALES

No member of these orders has been raised in pure culture. Unialgal cultures of species of *Vaucheria* can be grown from zoospores, except for *V. terrestris* which never forms swarmers. The zoospores, unlike the filaments, can be freed from other algae by washing, but in spite of numerous attempts I failed to eliminate bacteria. The period of swarming is so short and so easily interfered with by changes in surrounding conditions that it is impossible to pass the zoospores through several changes of sterile medium. They soon cease swarming and secrete a firm surface-layer which is immediately infected by bacteria.

CONJUGALES

Members of Mesotaeniaceae, Desmidiaceae and Zygnemaceae have been grown in pure cultures. This has been possible, in spite of the large size of the cells and the absence of zoospores, because the surface is free from bacteria so long as healthy growth continues. So far as is known, material from the natural habitat, provided it is young, is most suitable for the preparation of pure cultures. Which cells are actually free from bacteria can only be determined by plating. Conjugales are delicate, so that the agar should be well washed and rinsed several times with glass-distilled water. The mineral solution should be of a rather low concentration. The medium used by Czurda (1926) for *Spirogyra* contains: KNO_3 0·02 %, K_2HPO_4 0·002 %, $MgSO_4.7H_2O$ 0·001 %, $FeSO_4.7H_2O$ 0·0005 %, $CaSO_4$ 0·2 % of the saturated solution, pH 6. The agar medium should not be heated longer than is absolutely necessary, and should have only a very faint bluish, not a yellowish, tint when ready for use (cf. p. 51).

Single cells of Mesotaeniaceae and Desmidiaceae are easily picked up with a capillary pipette, while threads of Desmidiaceae and Zygnemaceae are handled with glass needles. Success in obtaining pure cultures is more frequently achieved if the plating is carried out immediately in the natural habitat, because bacteria multiply rapidly in the containers in which the material is conveyed to the laboratory (cf., for Zygnemaceae, Czurda, 1926, 1930; for Cosmarium, Ondraček, 1936). The Mesotaeniaceae and Desmidiaceae found in bogs are adapted to a high H·-ion concentration, but some, like many of the Zygnemaceae, live in eutrophic neutral or alkaline waters.

Pure cultures, when once obtained, are easily maintained because of the longevity of the cells. They grow well in solutions and especially in dilute agar media (cf. p. 54). All or at least most Conjugales seem to be strictly holophytic.

HETEROKONTAE

Species belonging to this class are cultivated in the same way as many Chlorophyceae. Zoospores or autospores should again be used for plating. Some species are adapted, or can be accustomed, to acid media, which makes it easier to reduce bacteria.

Vischer (1937) states that in cultures Heterokontae cannot be distinguished from Chlorophyceae by their colour, which in nature is commonly of a yellowish green tint. The species hitherto grown in pure culture are indeed usually quite as green as green algae, although they often become paler when the medium begins to be exhausted. The blue colour exhibited after treatment with hydrochloric acid is also not reliable, since it is not always realized. On the other hand Chlorophyceae, which become yellowish in a medium lacking nitrogen or iron, give the same reaction owing to the relatively high carotene content which they then possess.

A more reliable character for distinguishing Heterokontae and Chlorophyceae lies in the usual formation of starch by the latter, whereas it is always lacking in the former. To obtain certainty on this point a medium containing sugar is used, on which starch is quickly formed by all species which are able to synthesize it. Another distinguishing feature lies in the flagella, which are best investigated by treating a hanging drop containing zoospores with iodine vapour.

No other group of algae has so far been grown in pure culture.

REFERENCES

ALLEN, E. J. (1914). On the culture of the plankton diatom *Thalassiosira gracida* Cleve in artificial sea water. *J. Mar. Biol. Ass. U.K.* **10**, 417.

ALLEN, E. J. AND NELSON, E. W. (1910). On the artificial culture of marine plankton organisms. *J. Mar. Biol. Ass. U.K.* **8**, 421.

ALLISON, F. E. AND HOOVER, S. R. (1935). Conditions which favour nitrogen fixation by blue-green algae. *Trans. 3rd Int. Congr. Soil Sci.* **1**, 145.

ALLISON, F. E., HOOVER, S. R. AND MORRIS, H. J. (1937). Physiological studies with the nitrogen-fixing alga *Nostoc muscorum*. *Bot. Gaz.* **98**, 433.

ALLISON, F. E. AND MORRIS, H. J. (1929). Nitrogen fixation by blue-green algae. *Soil Sci.* **71**, 221.

BACHRACH, E. (1927). Quelques observations sur la biologie des Diatomées. *C.R. Soc. Biol., Paris*, **47**, 689.

BACHRACH, E. AND LEFÈVRE, M. (1930–1). Recherches sur la culture des Péridiniens. *Rev. algol.* **5**, 55.

BAKER, W. B. (1926). Studies in the life history of *Euglena*. I. *Euglena agilis* Carter. *Biol. Bull. Woods Hole*, **51**, 321.

BARKER, H. A. (1935). The culture and physiology of marine dinoflagellates. *Arch. Mikrobiol.* **6**, 157.

BEIJERINCK, M. W. (1890). Kulturversuche mit Zoochlorellen, Lichenogonidien und anderen niederen Algen. *Bot. Ztg*, **48**, 725.

BEIJERINCK, M. W. (1891). Verfahren zum Nachweis der Säureabsonderung bei Mikrobien. *Zbl. Bakt.* **9**, 781.

BEIJERINCK, M. W. (1893). Bericht über meine Kulturen niederer Algen auf Nährgelatine. *Zbl. Bakt.* **13**, 368.

BEIJERINCK, M. W. (1898). Notiz über *Pleurococcus vulgaris*. *Zbl. Bakt.* II, **4**, 785.

BEIJERINCK, M. W. (1901). Über oligonitrophile Mikroben. *Zbl. Bakt.* II, **7**, 561.

BEIJERINCK, M. W. (1904). Das Assimilationsprodukt der Kohlensäure in den Chromatophoren der Diatomeen. *Rec. trav. bot. néerl.* **1**, 28.

BENECKE, W. (1898). Über Kulturbedingungen einiger Algen. *Bot. Ztg*, **56**, 83.

BOLD, H. C. (1942). The cultivation of algae. *Bot. Rev.* **8**, 69.

BORTELS, H. (1940). Über die Bedeutung des Molybdaens für stickstoffbindende Nostoccaceen. *Arch. Mikrobiol.* **11** (2), 155.

BOUILHAC, R. (1897). Sur la culture de *Nostoc punctiforme* en présence de glucose. *C.R. Acad. Sci., Paris,* **125**, 880.

BOUILHAC, R. (1898). Présence de chlorophylle dans un *Nostoc* cultivé à l'abri de la lumière. *C.R. Acad. Sci., Paris,* **127**, 119.

BRISTOL-ROACH, M. B. (1926). On the relation of certain soil algae to some soluble carbon compounds. *Ann. Bot., Lond.,* **40**, 149.

BRISTOL-ROACH, M. B. (1927). On the carbon nutrition of some algae isolated from soil. *Ann. Bot., Lond.,* **41**, 509.

CALKINS, G. N. AND SUMMERS, FR. M. (1941). *Protozoa in Biological Research.* New York.

CHODAT, R. (1900). Cf. Chodat and Grintzesco.

*CHODAT, R. (1904). Culture pure d'Algues vertes, de Cyanophycées et de Diatomacées. *Arch. Sci. Phys. Nat.* **65**.

CHODAT, R. (1909). *Étude critique et expérimentale sur le polymorphisme des algues.* Genève.

CHODAT, R. (1913). Monographie d'algues en culture pure. Berne.

CHODAT, R. AND GRINTZESCO, J. (1900). Sur les méthodes de culture pure des Algues vertes. *Congr. Int. Bot., Paris,* p. 157.

CHU, S. P. (1942). The influence of the mineral composition of the medium on the growth of planktonic algae. I. Methods and cultural media. *J. Ecol.* **30**, 284.

CLEVELAND, L. R. (1928). The separation of a *Trichomonas* of man from bacteria. *Amer. J. Hyg.* **8**, 256.

CZURDA, V. (1926). Die Reinkultur von Conjugaten. *Arch. Protistenk.* **53**, 215; **54**, 355.

CZURDA, V. (1930). Experimentelle Untersuchungen über die Sexualitätsverhältnisse der Zygnemales. *Beih. Bot. Centralbl.* **47**, I, 15.

CZURDA, V. (1933). Experimentelle Analyse der kopulationsauslösenden Bedingungen bei Mikroorganismen. *Beih. Bot. Zbl.* **51**, 711.

DE, P. K. (1939). The role of blue-green algae in nitrogen fixation in rice-fields. *Proc. Roy. Soc. B,* **127**, 121.

DIWALD, K. (1938). Die ungeschlechtliche und geschlechtliche Fortpflanzung von *Glenodinium lubiniensiforme* sp.nov. *Flora, Jena,* **132**, 174.

* Papers marked with an asterisk have not been seen.

DUSI, H. (1930 a). Les limites de la concentration en ion H pour la culture d'*Euglena gracilis* Klebs. *C.R. Soc. Biol., Paris*, **103**, 1184.

DUSI, H. (1930 b). Les limites de la concentration en ion H pour la culture de quelques Euglènes. *C.R. Soc. Biol., Paris*, **104**, 734.

DUSI, H. (1931). L'assimilation des acides aminés par quelques Eugléniens. *C.R. Soc. Biol., Paris*, **107**, 1232.

DUSI, H. (1933). Recherches sur la nutrition de quelques Euglènes. I. *Euglena gracilis. Ann. Inst. Pasteur*, **50**, 550.

FAMINTZIN, A. (1871). Die anorganischen Salze als ausgezeichnetes Hilfsmittel zum Studium der Entwicklung niederer chlorophyllhaltiger Organismen. *Bull. Acad. Sci. St-Petersb.* **17**, 31.

FOGG, G. E. (1942). Studies on nitrogen fixation by blue-green algae. I. Nitrogen fixation by *Anabaena cylindrica* Lemm. *J. Exp. Biol.* **19**, 78.

FØYN, B. (1934). Lebenscyklus. Cytologie und Sexualität der Chlorophycee *Cladophora Suhriana* Kützing. *Arch. Protistenk.* **83**, 1.

FREUND, H. (1908). Neue Versuche über die Wirkungen der Aussenwelt auf die ungeschlechtliche Fortpflanzung der Algen. *Flora, Jena*, **98**, 41.

FRITSCH, F. E. (1935). *The Structure and Reproduction of the Algae*, 1. Cambridge.

FRITSCH, F. E. AND JOHN, R. P. (1942). An ecological and taxonomic study of the algae of British soils. II. Consideration of the species observed. *Ann. Bot., Lond.*, **6**, 371.

GEITLER, L. (1932). Der Formwechsel der pennaten Diatomeen (Kieselalgen). *Arch. Protistenk.* **78**, 1.

GERNECK, R. (1907). Zur Kenntnis niederer Chlorophyceen. *Beih. bot. Zbl.* Abt. 2, **21**, 221.

GLADE, R. (1914). Zur Kenntnis der Gattung *Cylindrospermum. Beitr. Biol. Pfl.* **12**, 295.

*GLASER, R. W. AND CORIA, N. A. (1930). Methods for the pure culture of certain Protozoa. *J. Exp. Med.* **51**, 787.

GUNDERSON, M. F. AND SKINNER, C. E. (1932). Suggestions for growing mass cultures of algae for vitamin and other physiological study. *Plant Physiol.* **7**, 539.

HAEMMERLING, J. (1931). Entwicklung und Formbildungsvermögen von *Acetabularia mediterranea. Biol. Zbl.* **51**, 633.

HAEMMERLING, J. (1934). Über die Geschlechtsverhältnisse von *Acetabularia mediterranea* und *Acetabularia Wettsteinii. Arch. Protistenk.* **83**, 57.

HALL, R. P. (1931). On certain culture reactions of *Euglena*. *Anat. Rec.* **51** (abstracts), 83.

HALL, R. P. (1939). The trophic nature of the plant-like flagellates. *Quart. Rev. Biol.* **14**, 1.

HALL, R. P. AND SCHÖNBORN, H. W. (1939). Selective effects of inorganic culture media on bacteria-free strains of *Euglena*. *Arch. Protistenk.* **93**, 72.

HARTMANN, M. (1921). Die dauernd agame Zucht von *Eudorina elegans*, experimentelle Beiträge zum Befruchtungs- und Todproblem. *Arch. Protistenk.* **43**, 223.

HARVEY, H. W. (1933). On the rate of diatom growth. *J. Mar. Biol. Ass. U.K.* **19**, 253.

HILDEBRAND, E. M. (1938). Technique for the isolation of single micro-organisms. *Bot. Rev.* **4**, 628.

HÖLL, K. (1928). *Oekologie der Peridineen.* Jena.

HUTNER, S. H. (1936-7). The nutritional requirements of two species of *Euglena*. *Arch. Protistenk.* **88**, 93.

JAAG, O. (1929). Recherches expérimentales sur les gonidies des lichens appartenants aux genres *Parmelia* et *Cladonia*. *Bull. Soc. bot. Genève*, II, **21**, 1.

JAAG, O. (1931). Morphologische und physiologische Untersuchungen über die zur Gattung *Coccomyxa* gehörenden Flechtengonidien. *Verh. schweiz. naturf. Ges.* **112**, 331.

JAAG, O. (1933 *a*). Über die Verwendbarkeit der Gonidienalgen in der Flechtensystematik. *Ber. Schweiz. Bot. Ges.* **42**, 724.

JAAG, O. (1933 *b*). *Coccomyxa* Schmidle. Monographie einer Algengattung. *Beitr. Kryptogamenflora der Schweiz.*

JACOBSEN, H. C. (1910). Kulturversuche mit einigen niederen Volvocaceen. *Z. Bot.* **2**, 145.

JAHN, T. L. (1931). The effects of hydrogen-ion concentration on the growth of *Euglena gracilis* Klebs. *Biol. Bull. Woods Hole*, **61**, 387.

JOHN, R. P. (1942). An ecological and taxonomic study of the algae of British soils. I. The distribution of the surface-growing algae. *Ann. Bot., Lond.*, N.S. **6**, 323.

KLEBS, G. (1896). *Die Bedingungen der Fortpflanzung bei einigen Algen und Pilzen.* Jena.

KÖHLER-WIEDER, R. (1937). Ein Beitrag zur Kenntnis der Kernteilung der Peridineen. *Öst. bot. Z.* **86**, 199.

KOSTYTSCHEW, S. (1926). *Lehrbuch der Pflanzenphysiologie.* Berlin.

KRICHENBAUER, H. (1937-8). Beiträge zur Kenntnis der Morphologie und Entwicklungsgeschichte der Gattungen *Euglena* und *Phacus*. *Arch. Protistenk.* **90**, 88.

KUFFERATH, H. (1930). *La culture des Algues*. Paris: Publication de la Revue Algolog.

KÜSTER, E. (1907). *Anleitung zur Kultur der Mikroorganismen*. 1. Aufl. Leipzig u. Berlin. (1913). 2. Aufl.

KÜSTER, E. (1908). Eine kultivierbare Peridinee. *Arch. Protistenk.* **11**, 351.

LEMMERMANN, E. (1910). *Kryptogamenflora der Mark Brandenburg, etc., Algen*. Leipzig.

LEMMERMANN, E. (1913). Flagellaten, **2**. Pascher's *Süsswasserflora*. Jena.

LIMBERGER, A. (1918). Über die Reinkultur der *Zoochlorella* aus *Euspongilla lacustris* und *Castrada viridis*. *S.B. Akad. Wiss. Wien*, Math.-Nat. Kl., Abt. 1, **127**, 395.

LINDEMANN, E. (1929). Experimentelle Studien über die Fortpflanzungserscheinungen der Süsswasserperidineen auf Grund von Reinkulturen. *Arch. Protistenk.* **68**, 1.

LOEFER, J. B. (1934). The trophic nature of *Chlorogonium* and *Chilomonas*. *Biol. Bull. Woods Hole*, **66**, 1.

LOEFER, J. B. (1936). Isolation and growth characteristics of the ' *Zoochlorella* ' of *Paramaecium bursaria*. *Amer. Nat.* **70**, 184.

LUND, J. W. G. (1942). The marginal algae of certain ponds, with special reference to the bottom deposits. *J. Ecol.* **30**, 247.

LWOFF, A. (1923). Sur la nutrition des Infusoires. *C.R. Acad. Sci., Paris*, **176**, 928.

LWOFF, A. (1929). La nutrition de *Polytoma uvella* Ehrenberg et le pouvoir de synthèse des protistes hétérotrophes. Les protistes mésotrophes. *C.R. Acad. Sci., Paris*, **188**, 114.

LWOFF, A. (1932). Recherches biochimiques sur la nutrition des Protozoaires. *Monogr. Inst. Pasteur*, Paris.

MAINX, F. (1927 *a*). Untersuchungen über Ernährung und Zellteilung bei *Eremosphaera viridis* de Bary. *Arch. Protistenk.* **57**, 1.

MAINX, F. (1927 *b*). Beiträge zur Morphologie und Physiologie der Eugleninen. I, II. *Arch. Protistenk.* **60**, 305, 355.

MAINX, F. (1929 *a*). Untersuchungen über den Einfluss von Aussenfaktoren auf die phototaktische Stimmung. *Arch. Protistenk.* **68**, 105.

MAINX, F. (1929 *b*). Biologie der Algen. *Tabul. biol., Berl.*, **5**, 1.

MAINX, F. (1931). Physiologische und genetische Untersuchungen an Oedogonien. 1. Mitt. *Zt. Bot.* **24**, 481.

MARSHALL WARD, H. (1899). Some methods for use in the culture of algae. *Ann. Bot., Lond.*, **13**, 563.

MATRUCHOT, L. AND MOLLIARD, M. (1902). Variation de structure d'une algue verte sous l'influence du milieu nutritif. *Rev. gén. Bot.* **14**, 113.

MEINHOLD, TH. (1911). Beiträge zur Physiologie der Diatomeen. *Beitr. Biol. Pfl.* **10**, 353.

MEYER, H. (1897). Untersuchungen über einige Flagellaten. *Rev. Suisse Zool.* **5**, 43.

*MIQUEL, P. (1890–2). De la culture artificielle des Diatomées. *Le Diatomiste*, **8**.

*MIQUEL, P. (1892). De la culture artificielle des Diatomées. *C.R. Acad. Sci.*, Paris, **94**, 780.

*MIQUEL, P. (1892, 1893). Recherches expérimentales sur la physiologie, la morphologie et la pathologie des Diatomées. *Ann. Microgr.* **4**, 273; **5**, 437.

MOLISCH, H. (1895). *Die Ernährung der Algen.* I. Süsswasseralgen. *S.B. Akad. Wiss. Wien*, Math.-Nat. Kl., Abt. 1, **104**, 783.

MOLISCH, H. (1896). *Die Ernährung der Algen.* II. Süsswasseralgen. *S.B. Akad. Wiss. Wien*, Math.-Nat. Kl., Abt. 1, **105**, 633.

MOLLIARD, M. (1925). *Nutrition de la Plante.* Paris.

NAEGELI, C. (1893). Über oligodynamische Erscheinungen in lebenden Zellen. *Denkschr. Schweiz. naturf. Ges.* **33**.

NEEDHAM, J. G., et al. (1937). *Culture Methods for Invertebrate Animals.* Ithaca, New York.

ONDRAČEK, K. (1935). Über die Brauchbarkeit einiger Glassorten für Algenreinkulturen. *Arch. Mikrobiol.* **6**, 532.

ONDRAČEK, K. (1936). Experimentelle Untersuchungen über die Variabilität einiger Desmidiaceen. *Planta*, **26**, 226.

PASCHER, A. (1913) and later. *Die Süsswasserflora Deutschlands, Oesterreichs und der Schweiz.* Jena.

PRINGSHEIM, E. G. (1912). Die Kultur von Algen in Agar. *Beitr. Biol. Pfl.* **11**, 305.

PRINGSHEIM, E. G. (1913). Zur Physiologie der *Euglena gracilis*. *Beitr. Biol. Pfl.* **12**, 1.

PRINGSHEIM, E. G. (1914 a). Zur Physiologie der Schizophyceen *Beitr. Biol. Pfl.* **12**, 49.

PRINGSHEIM, E. G. (1914 b). Die Ernährung von *Haematococcus pluvialis* Flot. *Beitr. Biol. Pfl.* **12**, 413.

PRINGSHEIM, E. G. (1918). Die Kultur der Desmidiaceen. *Ber. dtsch. bot. Ges.* **36**, 482.

PRINGSHEIM, E. G. (1920). Zur Physiologie von *Polytoma uvella*. *Ber. dtsch. bot. Ges.* **38** (8).

PRINGSHEIM, E. G. (1921 a). Algenkultur. Abderhalden's *Handb. biol. Arb. Meth.* Abt. XI, Heft 2, p. 377.

PRINGSHEIM, E. G. (1921 b). Zur Physiologie saprophytischer Flagellaten (*Polytoma, Astasia* und *Chilomonas*). *Beitr. allg. Bot.* **2**, 88.

PRINGSHEIM, E. G. (1921 c). Physiologische Studien an Moosen. I. Die Reinkultur von *Leptobryum piriforme* (L.) Schpr. *Jb. wiss. Bot.* **60**, 499.

PRINGSHEIM, E. G. (1926 a). Methoden und Erfahrungen. *Beitr. Biol. Pfl.* **14**, 283.

PRINGSHEIM, E. G. (1926 b). Über das Ca-Bedürfnis einiger Algen. *Planta*, **2**, 555.

PRINGSHEIM, E. G. (1930). Die Kultur von *Micrasterias* und *Volvox*. *Arch. Protistenk.* **72**, 1.

PRINGSHEIM, E. G. (1934 a). Über die pH-Grenzen einiger saprophytischer Flagellaten. *Naturwissenschaften*, **22**, 510.

PRINGSHEIM, E. G. (1934 b). Über Oxythrophie bei *Chlorogonium*. *Planta*, **22**, 146.

PRINGSHEIM, E. G. (1935). Über Azetatflagellaten. *Naturwissenschaften*, **23**, 110.

PRINGSHEIM, E. G. (1936 a). Das Rätsel der Erdabkochung. *Beih. bot. Zbl.* **55** (A), 100.

PRINGSHEIM, E. G. (1936 b). Zur Kenntnis saprotropher Algen und Flagellaten. I. Über Anhäufungskulturen polysaprober Flagellaten. *Arch. Protistenk.* **87**, 43.

PRINGSHEIM, E. G. (1937). Beiträge zur Physiologie saprophytischer Algen und Flagellaten. I. *Chlorogonium* und *Hyalogonium*. *Planta*, **26**, 631.

PRINGSHEIM, E. G. (1942). Contributions to our knowledge of saprophytic algae and flagellata. III. *Astasia, Distigma, Menoidium* and *Rhabdomonas*. *New Phytol.* **41**, 171.

PRINGSHEIM, E. G. AND MAINX, F. (1926). Untersuchungen an *Polytoma uvella* Ehrb., insbesondere über Beziehungen zwischen chemotaktischer Reizwirkung und chemischer Konstitution. *Planta*, **1**, 583.

PROVASOLI, L. (1937-8). Studi sulla nutrizione dei Protozoi. *Boll. Lab. Zool. agr. Bachic. Milano*, **8**, 3.

REICHARDT, A. (1927). Beiträge zur Zytologie der Protisten. *Arch. Protistenk.* **59**, 301.

RICHTER, O. (1903). Reinkultur von Diatomeen. *Ber. dtsch. bot. Ges.* **21**, 493.

RICHTER, O. (1906). Zur Physiologie der Diatomeen. *S.B. Akad. Wiss. Wien*, Abt. I, **115**, 27.

RICHTER, O. (1909). Zur Physiologie der Diatomeen. II. Die Biologie der *Nitzschia putrida* Benecke. *Denkschr. Akad. Wiss. Wien*, Math.-Nat. Kl. **84**, 657.

RICHTER, O. (1911). *Die Ernährung der Algen.* Leipzig.

SCHRAMM, J. R. (1914). Some pure culture methods in the algae. *Ann. Mo. Bot. Gdn*, **1**, 23.

SCHREIBER, E. (1925). Zur Kenntnis der Physiologie und Sexualität höherer Volvocales. *Z. Bot.* **17**, 336.

SCHREIBER, E. (1928). Die Reinkultur von marinem Phytoplankton und deren Bedeutung für die Erforschung der Produktionsfähigkeit des Meerwassers. *Wiss. Meeresuntersuch.* Abt. Helgoland, N.F. **16** (10), 1.

SCHREIBER, E. (1931). Über Reinkulturversuche und experimentelle Auxosporenbildung bei *Melosira nummuloides. Arch. Protistenk.* **73**, 331.

SCHROPP, W. AND SCHARRER, K. (1933). Wasserkulturversuche mit der A-Z—Lösung nach Hoagland. *Jb. wiss. Bot.* **78**, 544.

SKINNER, C. E. (1932). Isolation in pure culture of green algae from soil by a simple technique. *Plant Physiol.* **7**, 533.

SMITH, G. M. (1916). A monograph of the algal genus *Scenedesmus* based upon pure culture methods. *Trans. Wisc. Acad. Sci.* **18**, 422.

STREHLOW, K. (1929). Über Sexualität einiger Volvocales. *Z. Bot.* **21**, 625.

TERNETZ, C. (1912). Beiträge zur Morphologie und Physiologie der *Euglena gracilis* Klebs. *Jb. wiss. Bot.* **51**, 433.

TISCHUTKIN, N. (1897). Über Agar-Agarkulturen einiger Algen und Amoeben. *Zbl. Bakt.* **3**, 183.

TREBOUX, O. (1905). Organische Säuren als Kohlenstoffquelle bei Algen. *Ber. dtsch. bot. Ges.* **23**, 432.

*TRELEASE, S. F. AND PAULINO, P. (1920). The effect on the growth of rice of the addition of ammonia and nitrate salts to soil cultures. *Philipp. Agric.* **8**, 293 (cited by Trelease and Trelease, 1935).

TRELEASE, S. F. AND TRELEASE, H. F. (1935). Changes in hydrogen-ion concentration of culture solutions containing nitrate and ammonium nitrogen. *Amer. J. Bot.* **22**, 520.

USPENSKI, E. E. (1925). Iron as a factor in the distribution of algae (Russian). *Mem. Bot. Inst. Ass. Res. Inst. Phys.* etc. State University, Moscow.

USPENSKI, E. E. (1927). Eisen als Factor für die Verbreitung niederer Wasserpflanzen. *Pflanzenforschung.* Jena.

USPENSKI, E. E. AND USPENSKAJA, W. J. (1925). Reinkultur und ungeschlechtliche Fortpflanzung des *Volvox minor* und *Volvox globator* in einer synthetischen Nährlösung. *Z. Bot.* **17**, 273.

VISCHER, W. (1920). Études d'algologie expérimentale. Formation des stades unicellulaires, cénobiaux et pluricellulaires chez les genres *Chlamydomonas, Scenedesmus, Coelastrum, Stichococcus* et *Pseudendoclonium. Bull. Soc. bot. Genève*, sér. 2, **18**, 24.

VISCHER, W. (1937). Die Kultur der Heterokonten. Rabenhorst's *Kryptogamenflora*, **11**, 190.

WARBURG, O. (1919). Über die Geschwindigkeit der photochemischen Kohlensäurezersetzung in lebenden Zellen. *Biochem. Z.* **100**, 230.

WARÉN, H. (1920). Reinkulturen von Flechtengonidien. *Oefvers. Finska Vet. Soc. Förhandl.* **61**, No. 14.

WARÉN, H. (1926). Nahrungsphysiologische Versuche an *Micrasterias rotata. Comment. biol., Helsingf.*, **2**, Nr. 8.

WARÉN, H. (1933). Über die Rolle des Calciums im Leben der Zelle auf Grund von Versuchen an *Micrasterias. Planta*, **19**, 1.

WARIS (WARÉN), H. (1936). Über das Calciumbedürfnis der niederen Algen. *Planta*, **25**, 315.

WARIS (WARÉN), H. (1939). Über den Antagonismus von Wasserstoffionen und Metallkationen bei *Micrasterias. Acta bot. fenn.* **24**, 3.

WETTSTEIN, F. v. (1921). Zur Bedeutung und Technik der Reinkultur für Systematik und Floristik der Algen. *Öst. bot. Z.* **70**, 23.

WHITE, P. R. (1938). Accessory salts in the nutrition of excised tomato roots. *Plant Physiol.* **13**, 391.

WIEDLING, ST (1941). Cultivation of *Nitzschia. Bot. Notiser*, p. 37.

WINOGRADSKY, S. (1896). Kulturversuche mit Amoeben auf festem Substrat. *Zbl. Bakt.* **19**, 257.

ZUMSTEIN, H. (1900). Zur Morphologie und Physiologie der *Euglena gracilis* Klebs. *Jb. wiss. Bot.* **34**, 149.

INDEX

Printed in the United States
By Bookmasters